❈ The Lost Cement Mine ❈

The Lost

PUMICE MOUNTAIN . . . *the truly grand dome of . . . Pumice Mountain, a great, round mass of yellowish white lava and pumice stone, ribbed with red streaks.*

Cement Mine

James W.A. Wright

Edited by Genny Smith

Illustrated by Nina Kelley

Mammoth Lakes, California
Genny Smith Books

Previously published as *The Cement Hunters*
by Dawson's Book Shop, Los Angeles, 1960
Edited by Richard E. Lingenfelter

Cover and book design for this edition by David Mike Hamilton
Copyright © 1984 by Genny Smith

Library of Congress Cataloging in Publication

Wright, James W. A. (James William Abert), 1834–1894.
The lost cement mine.

Reprint with some new material. Originally published:
The cement hunters. Los Angeles: G. Dawson, 1960.
Bibliography: p.
1. Gold mines and mining—California—Owens River Valley—
History. 2. Gold mines and mining—California—Owens River
Valley—Folklore. 3. Legends—California—Owens River Valley.
4. Owens River Valley (Calif.)—History, Local. 5. Treasure-
trove—California—Owens River Valley.   I. Smith, Genny
Schumacher. II. Title.
F868.09W75 1984 979.4′48 84-7991
ISBN 0-931378-08-7
ISBN 0-931378-09-5 p

Distributed by William Kaufmann, Inc.
95 First Street, Los Altos, California 94022

·→❧· **Contents** ·❧←·

Publisher's Preface to This Edition

Among hundred-year-old files of the San Francisco *Daily Evening Post*, Richard Lingenfelter discovered this story of the Lost Cement Mine by James W. A. Wright. Wright wrote his account in the fall of 1879, after a summer visit to Mono County's mining camps in eastern California.

Wright's report is partly hearsay, partly a day-by-day account of the country he traveled between Monoville and Mammoth City. So detailed are his descriptions that you can locate today many of the places he wrote about more than a hundred years ago. For example, you can easily follow his route up Red Mountain, south of Mammoth City, to the top of his "highest gray granite peak." There you can enjoy the same magnificent views that enthralled him on the eighth of July, 1879, and that he wrote about at length in the Introductory.

Fortunately for us, Lingenfelter rescued this story from oblivion, and in 1960 Dawson's Book Shop published it in a limited edition of 200 copies under the title *The Cement Hunters*. This 1984 edition reproduces the text type of the Dawson book, although we have re-set the preliminary pages. If some spelling and punctuation seems odd, be assured that we have reproduced Wright's text exactly as it appeared in the *Post*. New material in this edition consists of: three historic maps, additional notes, Mark Twain's full account of the Cement Mine and Nina Kelley's illustrations. Kelley's drawings accurately portray this eastern Sierra region that is the setting for Wright's story; landmarks are clearly recognizable.

Perhaps, one day, somewhere on the headwaters of the Owens River, you too may hear the tap! tap! tap! of a prospector's hammer, just as Wright did over a hundred years ago. Don't be surprised if it comes from a cement hunter. I know of at least one who, last summer, was searching still for that elusive ledge with flakes of purest gold.

ACKNOWLEDGMENTS

My grateful appreciation to the following institutions and individuals for their generous cooperation. The Bancroft Library and Annegret Ogden. Stanford University Libraries and Carol Bickler, Special Collections; also Karyl Tonge, Central Map Collection. U.S. Geological Survey Library, Menlo Park, and Eleanore Wilkins and Bill Sanders. Robert O. Davis, retired, and N. King Huber, U.S.G.S.; Jean DeMouthe, California Academy of Sciences; Demetrius Pohl, Stanford School of Earth Sciences. Special thanks to Richard Lingenfelter and Dawson's Book Shop for permission to reprint *The Cement Hunters*, and very special thanks to Mike Hamilton for designing and producing this handsome book.

Mammoth Lakes, July 1984   Genny Smith

-❧ Preface ❧-

For over a century now the desert country east of the Sierra has been the scene of countless quests for the elusive fortunes of "lost mines." One of the oldest and most sought after of these is the evasive gold ledge known as the "Lost Cement." Legend places this bonanza somewhere along the rugged eastern slope of the Sierra about the headwaters of the Owens River and describes it as a wide curbstone of reddish cement, richly bejeweled with nuggets of coarse gold.

Unlike the famous "Lost Gunsight," whose discovery by the Death Valley Forty Niners is well recorded in history, the original discovery of the "Lost Cement" is given to us only in the meagerest of legends. The most generally accepted version relates simply that the ledge was found by two men while on their way west across the Sierra in 1857. Needless to say, this legend finds no further corroboration. Yet, possibly even because of the vagueness surrounding its initial discovery, the "Lost Cement" has been the goal of numberless searches, which in themselves have formed a very real and important part of the history of the desert country.

The first quest for the "Lost Cement" began in 1861 and the search gained its widest attention the following decade, when it was colorfully set in print in Mark Twain's *Roughing It*.[1] By 1879 the search had led inadvertently to the discovery of many real and valuable lodes and the subsequent organization of a number of promising new mining districts. To these new discoveries came James W. A. Wright, a veteran mining correspondent of the San Francisco papers. Wright was soon intrigued by stories of the "Lost Cement" and, after talking with Gideon F. Whiteman and J. F. Wilson, both prominent participants in early searches, he commenced a brief history of the first two decades of Cement hunting. This history was first published in weekly serial form in the San Francisco *Daily Evening Post* between November 8 and December 13, 1879,[2] and portions of it were afterward copied in the Mammoth City, Bodie, and Independence papers on the eastern slope.

This new publicity, coupled with the fact that the local Paiutes persistently displayed nuggets of coarse gold whose source they would not reveal, further stimulated the Cement hunt. In the summer of 1880 a great many parties took to the back canyons and foothills at the headwaters of the Owens River. Some simply searched aimlessly while others sought an inside track by hiring Indian guides and one man even came up from Independence guided by a divine revelation granted him in a dream. None found the elusive ledge, however, and new parties took their places the following season, but their numbers gradually decreased with the passing of years.

In 1886, Al Cain, an Antelope Valley rancher, triumphantly announced that he had at last found the "Lost Cement" but on further exploration his find proved worthless. The Cement hunt continued. Revival of mining activity at Lundy a few years later also provided new Argonauts for the quest. This time a large, but abortive, expedition struck out for Lee Vining Creek on a clue culled from a Paiute buck with a sack of flour and a gallon of whiskey.

Still, no one ever found the "Lost Cement" again and with the turn of the century the searchers have grown fewer and fewer— and today there are nearly none. But in all, the legend of the "Lost Cement" has done well by many of those who sought it and it has done its part in the development of the eastern slope. Yet the Cement itself can never be found, for that is the way of "lost mines" and that is what makes them lost.

R.E.L.

1 Chapter XXXVII, *see* Appendix
2 As "The Owen's River War and The Cement Hunters"

·—❧ James ❦—·
William Abert Wright

When J. W. A. Wright came to Mammoth City in the summer
of 1879, he came both as a prospective investor and as a mining
correspondent of the San Francisco *Daily Evening Post* and the
Mining and Scientific Press. He spent several weeks on the eastern slope, visiting and appraising the prospects of the new mining
districts that crowded the Sierra from Mono Lake to Bishop
Creek. During this time he wrote a number of letters to both papers and collected the notes for the history of the Cement Hunters, which he was to write on his return to Hanford that fall.

Wright had been a correspondent to several newspapers for
nearly twenty years, but his interest in mining spanned only a
decade since his arrival in California. He was born July 28, 1834,
at Columbus, Mississippi, where his father, the Reverend David
Wright, was a Presbyterian missionary to the Choctaw Indians.
When he was only six years old, James' father died, leaving the
family nearly destitute, as all of their property—including several
slaves—had been lost in the crash of the "wildcat" banks a few
years earlier. Following his father's death, Wright's mother began
giving music lessons to support them and, aided by relatives, she
was able to help put him through college.

In January of 1853, Wright entered Professor Henry Tutwiler's
high school at Green Springs, Alabama. The following year he
served as assistant teacher at the school and thereby earned the
money needed to complete his college education. He entered the
junior class at Princeton University in August of 1855 and gradu-
ated in the class of '57 as valedictorian. That fall he returned to
Alabama and taught at Professor Tutwiler's high school until 1862.
The year after his return he married the professor's oldest daughter, Margaret.

At the beginning of the Civil War, Wright had voted against se-
cession, but in the spring of 1862 he accepted a commission in the
Confederate Army and raised a company of which he was elected
Captain. His command became Company H of the 36th Alabama

Infantry and served under Generals Bruckner, Bragg, Johnston, Hood and Taylor. Wright was severely wounded in the right hip at the battle of Missionary Ridge on November 25, 1863, and was taken prisoner. He was confined in hospitals at Chattanooga, the State Prison at Nashville, and Camp Chase near Columbus, Ohio. But on March 27, 1864, he made his escape from a prison train near Harrisburg, Pennsylvania, while enroute to Fort Delaware near Philadelphia. He contacted friends in Philadelphia, who helped him make his way into Canada and from there to Bermuda. He finally returned to Dixie on the blockade runner *Lilian*, landing at Wilmington, North Carolina, on June 4, 1864. After a brief leave with his family in Alabama, he returned to duty and for the last nine months of the war he acted as a field officer, during which time he was promoted to Major.

At the close of the war, Wright turned down an offer to go into a law practice and returned to his teaching position with Professor Tutwiler, remaining there until the spring of 1868, when he decided to move to California. Going by way of New York and Panama, he arrived at San Francisco aboard the steamer *Nevada* on June 13, 1868. He went at once to Stockton and joined a number of Southern friends, whom he accompanied to Fresno county, where they located the Alabama Settlement. Wright also purchased some farm land near Turlock and there he made his home, settling down to grow wheat and barley. He remained there for six years, during which time he became active in the Grange and was elected Master of the State Grange of California on its organization at Napa City.

In June of 1878, Wright moved to Hanford, Tulare county, where he was to remain for many years. In the late 1870's he began spending more and more of his time on trips to the Sierra and the new boom camps of California, Nevada, and Arizona. Within about four years he had explored a great part of the high Sierra south of Yosemite and in 1882 he made a series of maps of the country for W. W. Elliott's "Guide to the Scenery of the Sierra Nevada," "The History of Fresno County," and "The History of Tulare County." The Sierra Club in 1925 gave his name to Wright Lakes and Creek near Mount Whitney, which he had explored in 1881 in company with W. B. Wallace and F. H. Wales.

JAMES W. A. WRIGHT

J. W. A. Wright

The Lost Cement Mine

A legendary gold mine supposedly discovered in 1857 on the eastern slope of the Sierra Nevada, Mono County, California. Tales of "the burnt country" at the headwaters of the Owens River—a country of black lava and white pumice, and somewhere a ledge of reddish cement spangled with flakes of purest gold.

❧ MINERAL HILL ❧

Mineral Hill (now called Red or Gold Mountain), viewed across Twin Lakes from the south slope of Mammoth Mountain.

"Were the painter, the novelist, the tourist, or the geologist, naturalist or any other scientist, to search the world over for a point, easy of access, that combined most wonderfully . . . all elements of the sublime, immense, picturesque, curious, weird and varied in forms, colors and startling contrasts, and extent of mountains, valley and lake scenery, he would not be far wrong to locate it on one of the highest granite or lava peaks, near the new mining town of Mammoth City, Mono county, California. . . . A point suitable and easily ascended . . . is the highest gray granite peak . . . that rears its snow encircled head about four miles southeast of Mammoth City in a direct line across Mineral hill."

MINERAL HILL   *Center skyline*: the red cone Wright called Mount Parker. *Far Right*: "the highest gray granite peak" that he climbed on the 8th of July, 1879.

⚬➤ Introductory ➤⚬

LOCATION OF THE EVENTS TO BE RELATED

Were the painter, the novelist, the tourist, or the geologist, naturalist or any other scientist, to search the world over for a point, easy of access, that combined most wonderfully in the domains of nature all elements of the sublime, immense, picturesque, curious, weird and varied in forms, colors and startling contrasts, and extent of mountains, valley and lake scenery, he would not be far wrong to locate it on one of the highest granite or lava peaks, near the new mining town of Mammoth City, Mono county, California. In fact we may seriously question, after a fair comparison, whether any part of the world, accessible or inaccessible, more fully and attractively unites all features above indicated, where the confines of civilization and the contrary barely touch each other, than does that truly strange region among our "Higher Sierras."

MOUNTAIN CLIMBING.

A point suitable and easily ascended to prove this is the highest gray granite peak—nameless as yet—that rears its snow encircled head about four miles southeast of Mammoth City in a direct line across Mineral hill. This peak can readily be reached on foot, with the necessary halts for breathing in that rare atmosphere, in from four to six hours of rough mountain climbing, by taking the trail that leads from the upper tunnels and reddish outcroppings of Mammoth mine to the Headlight, its southern extension, the mouth of whose shaft is just about 11,000 feet above sea-level. Be sure to take a good solid lunch with you, for the bracing mountain air and the exertion will give you a ravenous appetite. If preferred, a winding but comparatively easy ascent can be made on horseback by following the wagon road to Pine City, nearly a mile southwest of Mammoth, and along three of the larger of the fine fresh-water lakes that form the sources of the Mammoth branch of Owen's river. Thence the road and trail lead you through the highest pine forest (*pinus contorta*) in that district, nearly 11,000 feet up. To the left of this, at an altitude quite 100 feet higher, you pass the loghouse of J. W. Furlong, where he and his wife enjoy a vastly higher life than any on Nob hill. Thence, turning to the southeast, you can ride if you choose to the top of a red cone, formed chiefly of shattered talcose slate and called

MOUNT PARKER,[1]

In honor of one of the four original discoverers of the Alpha, or first mineral ledge located in this Lake district, in 1877. Here the altitude is about 11,800 feet,[2] and the view that greets the eye is already extensive and truly magnificent. But, for the still more commanding point, ride down the southeastern slope, and by zigzag course, for your horse's sake—because there is no trail—you can ride to a long, reddish white ridge, nearly a mile farther, and there horses can be safely tied to some stunted but strong shrubs on the eastern slope, where the rocky crest will form a partial shelter from the sharp wind that rages there at almost all hours. Thence, you leave to the right two dark slate peaks, making your way with difficulty along their eastern slopes a few hundred feet below their tops, finding a treacherous footing on their sharp-edged debris. In the sag beyond, and just before commencing the next ascent, you pass a line most distinctly and curiously marking the limit where the slaty formation on the west unites with the

HUGE GRANITE MASSES

That form the eastern wall of this mineral belt. A bold peak, 12,400 feet above the sea, formed of immense broken masses of granite, almost fit for the builder, lies in your direct course southeast. But it is best to pass along the western and southern slope of this, and avoid the fatigue necessary for its ascent, because the higher peak we wish to reach is about a mile farther southeast. If you do not have to stop every hundred yards or two on this part of the route and. puff and blow till you can breathe with comfort, and if you do not already have in your mouth that peculiar copperish taste, forewarning you that the removal of the accustomed pressure of air by its lightness here, is allowing the blood to force its way towards the surface of the membranes of your lungs and mouth, you may congratulate yourself that you can make a fair mountain climber and can push ahead without apprehension. If you leave Mammoth about eight o'clock in the morning, you may hope to have reached your destination by about two P.M. After resting a few moments on this giddy hight, at least 13,000 feet above sea level, you will recover from the necessary strain on muscle and lungs. The snow along the route will furnish water for refreshment. Then look around, taking in every point of the compass, and you will feel richly repaid for all your toil in reaching this high summit, which commands for us so superb a view of a large part of

THE GRANDEST MOUNTAIN CHAIN OF OUR CONTINENT.

As your gaze sweeps rapidly along the distant horizon, and you take in at a glance the superb and wonderfully varied panorama unfolded for at least a hundred miles around, the multitude of cones, and domes, and pinnacles, and needles, and palisades, and ridges, and spurs, with every conceivable form of depressions, filled with masses of snow, and lakes, with tree clad and barren valleys far beneath you, the glorious view is almost bewildering. It is no exaggeration to say that you are surrounded by at least a thousand visible mountain tops of more or less prominence. Their varied coloring adds no little to the grandeur of this mountain scenery. Gray, crimson, lilac, red, yellow, pink, black and white are charmingly intermingled with the deep green of dense forests to westward, and the blue waters of a dozen lakes in different directions, while above is the deep blue sky without a cloud. Such, at least, was this vast, serene, impressive view when enjoyed by the writer on the 8th of July, 1879, in company with his kind hearted, intelligent guide, John Riordon, a pioneer miner of that district. Midsummer as it was, the mercury at 2:30 P.M. in that high region showed a temperature of 53 degrees Fahrenheit in the scant shade we could form for it with our bodies, and the west wind was blowing a steady gale.

Immediately south of this lofty peak is a sheer precipice 2,000 feet or more along its almost perpendicular sides; then down, down steep slopes, southward as well as to the northeast, you see

OWEN'S RIVER

Winding along, from 5,000 to 8,000 feet below, in its course slightly east of south, to its sink in Owen's lake, about 80 miles distant in a direct line. From its sources, in a series of handsome lakes, mostly fresh water, along the steep eastern slopes of Mt. Lyell[3] and Mammoth mountain, its entire length is about 150 miles.

Just fifteen miles west of the north end of Owen's lake, and on the boundary line between Tulare and Inyo counties, is Mt. Whitney, 15,000 feet high, the culminating point of our highest Sierras. Along the eighty miles between it and your point of observation, and to the right and left of you, myriads of peaks 13,000 and 14,000 feet in altitude and upwards glitter in the sunlight with their white and gray and red crests and begin to cast their shadows to the eastward. You are here eighty-five miles, slightly east of north, from Visalia, seventy-five northeast of Fresno City, sixty-five northwest of Independence, while Walker's pass lies about 140 miles a little east of south from your position. San Francisco is nearly due west 190 miles. From here to 40 miles east of the great summit ridge of the Sierra Nevada, on a spur of which you stand, and forming the eastern boundary of Owen's valley for about 130 miles, is the long but lower range known in succession from north to south as the White, the Inyo and the Coso mountains. In a nearly due southeast line, crossing this ridge but a few miles north of the noted Pah Ute Monument, you look over the most remarkable depression of surface in North America, unless parts of the Colorado desert are to

⟶❈ VIEW WEST ❈⟵

FROM MINERAL HILL

"But the most charming, dazzling, altogether splendid and fascinating view, is toward the northwest, west, and southwest. There your eyes are feasted at one sweep with the wonderful grandeur of the Mount Lyell group with Dana north, Ritter and the Minarets south of it. . . . Nearly due west of us is Pumice gap, the pass now used for the trail from the valley of the San Joaquin. . . . Immediately north of this is the truly grand dome of Mammoth or Pumice mountain, a great, round mass of yellowish white lava and pumice stone, ribbed with red streaks."

VIEW WEST FROM MINERAL HILL. *Here the "great summit ridge" of the Sierra Nevada consists of, from left to right:* Mammoth Crest, Pumice Gap (Mammoth Pass), and Pumice or Mammoth Mountain. *Center, far distance:* the Minarets, Mount Ritter, Banner Peak. *Center:* Lake Mary.

be excepted. One hundred and twenty-five miles from you in that direction is

DEATH VALLEY,

Of which too little as yet is known and written, when we consider its remarkable nature. This strange depression is 65 or 70 miles long from northwest to southeast, and 5 or 15 miles wide. Its northern part is only five miles east of the crest of Panamint mountains. Yet the bottom of this dismal, mountain-locked valley was found by our state geological survey to be from 100 to 250 feet below the level of the Pacific ocean. This, too, when Mt. Whitney, our highest peak, with its lofty regions, is only 65 miles due west of its northern part. This valley is the sink of the comparatively unexplored, and hence mysterious, Amargosa river, which rises in western Nevada, about 40 miles north of Death valley, flows southeast 100 miles or more, thence west and northwest 40 or 50 miles farther, and empties into the southern end of this literal Dead sea of California. Yet at no time does this valley contain much water. It is merely marshy towards the center in winter. Its whole surface is covered with so thick and white an efflorescence of alkali, as to be almost blinding to one who long remains in it. But a long stay there is out of all question, for no fresh water is found in the valley, and it is doubtful whether any well could obtain it. This valley owes its gloomy name to the fact that a party of immigrants, on their way to California, perished there—from thirst, as is supposed—in 1849. Their wagon tracks, with remains of their fires and wagons, were not found until 11 years later, as is described by Dr. Owen—after whom Owen's river, valley and lake were named[4] —in his report, in April, 1861, to the United States Boundary Commission. An item of interest is that this odd depression below sea level—the only one any distance from an ocean in North America—is but 180 miles in a direct line northeast from the nearest point on the Pacific, in Santa Monica bay. Its north point is 270 miles due east of Monterey. About forty miles directly east of you, and near the line between California and Nevada, is the tall, white, sugar loaf cone called White Mountain peak,[5] supposed by Professor Whitney to be at least 14.600 feet high, with the possibility

suggested that it may be our highest peak. Far to the northward you see distinctly

THE EXCELSIOR, WASSUCK, AND SWEETWATER RANGES,

With all the barren, unattractive looking mountain regions
around Aurora, Bodie, Bridgeport and the old settlements of
Mono and Dogtown, their peaks uncovered by snow. There
you are overlooking the famous Walker's river country, about
100 miles away. Thirty miles due north of you the surface
of Mono lake glistens in the sunlight, divided into two parts
by the chain of Obsidian mountains extending ten or twelve
miles south of it, and containing, as does the queer lake itself,
the craters of extinct volcanoes which ages ago scattered
pumice and lava and small pink granite bowlders over a large
area north of you. But the most charming, dazzling, altogether splendid and fascinating view, is toward the northwest,
west, and southwest. There your eyes are feasted at one
sweep with the wonderful grandeur of the Mount Lyell group
with Dana north, Ritter and the Minarets south of it. Then
westward the Mount Clark or Obelisk group, and a vast
number of peaks of many forms up and down the San Joaquin
and its tributaries. For, you must remember that your great
altitude now enables you to look well over the huge ridge of
the Sierras, some two miles west of you, at every point except
nearly due south where very high jagged peaks rise above
your point of observation, and 3 or 4 miles from you. From
Whitney's description and map,[6] I take the highest of these
to be the

RED SLATE PEAK

Which he describes as eight miles north of the pass (12,400
feet high), where his party crossed from Round valley to the
San Joaquin. This peak they found to be 13,400 feet high,
which would place it at least 400 feet above our present point
of observation. Near us, to eastward, is the long, round
topped, purple, volcanic ridge in Laurel district, and now
known as Laurel mountain, that must certainly be not far
from 13,000 feet high. This, with the high region south of us,
is the home of the huge wild mountain sheep of California.
Nearly due west of us is Pumice gap, [7] the pass now used
for the trail from the valley of the San Joaquin, and which

we found to be at least 10,350 feet at the crest. Immediately
north of this is the truly grand dome of Mammoth or Pumice
mountain, a great, round mass of yellowish white lava and
pumice stone, ribbed with red streaks. Being immediately
between us and the Minarets, we can see that the estimate
of 13,000 feet for it by our geological survey is perhaps a few
hundred feet too much, for we, at about that altitude, easily
overlook its highest part, and the line of vision extends far
down along the base of the Minarets, going to prove that they
are certainly between 13,000 and 14,000 feet high. Of the
whole view, nothing is grander than the entire range to westward, the vast masses of perpetual snow, the numerous amphi-
theaters, the beautiful lakes, the rich forests, the numerous,
jagged, many-colored peaks. It reminds one forcibly of the
superb and world renowned

VIEW FROM RIGHI KULM, NEAR LUCERNE,

In Switzerland, as you look towards the lofty, snow-clad
Jungfrau and its attendant peaks. Yet, the California view,
perhaps, surpasses the Swiss in this. It has almost the same
elements of sublimity and beauty in the same degree, and to
this is added a more extended field of vision. The view from
this nameless gray peak commands a territory of unsurpassed
magnificence in the way of Alpine scenery, scarcely less than
200 miles from north to south and 100 from east to west.
This most remarkable mountain and valley region of California, in which we have sought to give our readers some insight
by a faint but truthful picture, is to southward and eastward
the scene of the Owen's river war which was waged fiercely
with the Mono Indians of that valley from 1861 to 1865,
contemporaneous with that greater war which we all so much
lament. To the northward especially occurred the thrilling
incidents in the persevering efforts of the "Cement Hunters,"
which, beginning in 1857, have scarcely ended to this day.
Future chapters will give an outline of these stirring events,
seeking to follow facts, as nearly as possible, and to give this
story the reliability of history.

❖ HEADWATERS ❖

OF THE OWENS RIVER

*"He gradually grew worse, until . . . he one day re-
quested the doctor to hand him a satchel that hung near
his bed. From it he took what was left of the lump of ore
that he had brought from the divide between Owen's
river and the San Joaquin. . . . In 1861, amid lively
times, quite a stir was created in this mining camp of
Mono by the arrival of a stranger, who was soon learned
to be a Dr. Randall, from San Francisco. . . . He engaged a man, and with him went out and located a
quarter section of land on 'Pumice flat,' some 37 miles
south of Mono, and about 8 miles north of where Mammoth City now stands."*

HEADWATERS OF THE OWENS RIVER The great summit ridge of the Sierra Nevada here divides the headwaters of the Owens River, on its eastern slope, from the headwaters of the San Joaquin River on its western slope. *Skyline, left of center*: Deadman Pass. *Far Right*: San Joaquin Mountain, Carson Peak. *Foreground*: Pumice Flat.

⋅→⁂ Original Story ⁂←⋅
of the Cement Hunters

Who that is at all acquainted, from personal experience or reading, with the mining developments of California, has not heard more or less of lost mines of rare richness, scattered here and there among the truly immense metal-bearing mountain chains of the Pacific coast? Many, many is the miner's yarn that has been spun about the details of these rich "finds," their loss, their present obscurity, and faith that they will be found again some day. These have served to while away the time many an evening, around the campfire; or in the miner's rude log cabin among our grand mountain fastnesses; or in the refined family circles of our ten thousand happy rural and city homes. Some of these wild stories are, doubtless, mere fabrications of the brain, while others are founded on facts. True or false, some of them have actually proved a great stimulus to thorough prospecting, by which really valuable mines have been found, though differing in character from those sought. In this they have served a good purpose, after all, and are worthy of record.

TO CITE A FEW INSTANCES:

There is the quaint old story about the lost "Blue Bucket" mine in the Malheur country, Oregon. In that case, a company of emigrants, in 1852, found in the mountains big nuggets of gold, ninety per cent in purity. These were abundant, but taking only a few with them, they had to hurry across the mountain wilderness, that they might not get out of supplies. Then they expected to return. How like a thousand human expectations! They left a blue wooden bucket near the spot, and where anybody could find that there was plenty of gold. Thousands have hunted for that bucket and the mine, yet none have found either. The search for them, however, has led to the discovery of the fine Virtue mine, and others in El Dorado district, Baker county and Grant county, on Snake river and Granite creek, Oregon. Then remember the "Lost Cabin" story, where a party built a log cabin in 1849 on the headwaters of Pit river, and near by were very rich placer diggings of coarse gold. That cabin and locality have never been found since, according to common repute, though some men are hunting for them still.

Again, among the most noted is the lost gold mine of Upper King's river, in the red slate belt, and called "Shipe's mine." It is known positively that a noted Indian fighter, a somewhat desperate man, named Shipe, did find such a mine of great value twelve years ago. He brought from it at least $600 worth of rich ore to Visalia, and that very night he was killed in a difficulty. His secret died with him, except that he had given a friend a very general description of the location of the ledge. Diligent search has been made, without finding it as yet.

There is also an account of some United States officers and men once hastening across the Sierras, and finding very rich gold deposits, never rediscovered, in the bed of a supposed tributary of the San Joaquin, Merced or Tuolumne.

Lately one of our city papers gave an interesting sketch of the tradition among the Indians of a "lost silver mine," possibly in Mono county, where ages ago Spaniards came every summer and smelted and packed away blocks of the white metal. Late discoveries there make this highly probable.

Here we shall give in detail the noted story of "the Cement Hunters" of Mono county, in its earliest form. Beyond question, it has played an important part in the rich discoveries of that region for three years past.

ORIGINAL FINDERS OF THE CEMENT.[8]

In the summer of 1857 a party of emigrants, on their way to the gold fields of California, by the southern route, reached Death valley, that dismal sink of Amargosa river, whose lowest points are nearly 300 feet below sea level, and which was briefly described in the first chapter. In braving its then unknown terrors and attempting to cross, they met with disaster such as too often overtook those who descended into this hot, parched, alkali-covered chasm. Nor can we wonder at its fatality and the painful sufferings of those pioneers who succeeded in crossing it, when we remember that a well equipped part of United States surveyors, who entered it in 1874, were glad to get out after 48 hours of most trying experience, by which they learned that the temperature of its close atmosphere did not in such seasons fall below 117 degrees Fahrenheit at midnight, while its day time heat and glare were so intense as almost to craze men and horses.

In its depths two men of this emigrant party lost their teams, and had to leave their wagons and outfits. They kept with the train, however, till they were nearing the eastern slope of the Sierra Nevada. When they reached a point in what is now Owen's valley, they were told by their comrades that it was only about sixty miles directly across these mountains to where gold digging was going on, but that the wagons must take a much longer route towards the north to find a suitable crossing. The two then left the train on foot, and started with their packs for the Nevadas. They came to the head of a stream, supposed to be Owen's river, and in traveling through "the burnt country" they sat down to rest.[9] Some say this was near a spring or stream. Here they observed a curious looking rock, which they commenced pounding. They saw in it a good deal of what appeared to be gold. There was so much, in fact, that they discussed the question whether it was gold or not. One insisted that it was, the other laughed at him. The one believing it was gold took about ten pounds of the rock with him. Reaching another stream, beyond the divide, they crossed these rough mountains successfully by following the river down, and it brought them out at

MILLERTON. [10]

Hence, this is known to have been the San Joaquin. There they found mining going on and tried their fortunes. After a time they separated. The one with the lump of ore was unsuccessful in the mines, and finally made his way to San Francisco.

A year or two passed. He was taken down with consumption, and placed himself for treatment under a Dr. Randall. [11] He gradually grew worse, until, when past all hope of recovery, he one day requested the doctor to hand him a satchel that hung near his bed. From it he took what was left of the lump of ore that he had brought from the divide between Owen's river and the San Joaquin.[12] He told the doctor he had learned with certainty at Millerton that it was gold, and had hoped to return some day to the place where he found it, and to enjoy the rich profits of the treasure which lay buried there. Now he knew he could not, and gave the remnant to the doctor, as all that he could pay him for his attentions. Yet, as a further return for his services, he gave him a minute description of the locality where he found it, saying it was on a stream heading in the Nevadas on the opposite side from the San Joaquin river. He also made a rough map for him, fixing the exact place, as nearly as he could, before his death.

This valuable secret was thus made the sole property of Dr. Randall in the fall of 1860. Having full faith in the earnest statement of a dying man, and having tangible and convincing evidence from the rich specimens he possessed, that such gold bearing cement did exist and was more than half gold, he quietly determined to tell no one, but to make thorough search for it the next year. This specimen is described by those who saw it afterwards, as a reddish, rusty looking cement,[13] not unlike decomposed quartz, and thickly spangled with flakes of purest gold. Meanwhile,

THE FIRST SETTLEMENT

By white men near Mono lake had been made in 1857, on Virginia creek, the east fork of Walker's river. This was called Dogtown. The next spring, all hands, except John Richter, moved thence to where the new town of Mono [14] was located, on the north side of Mono lake, and as regards the present flourishing towns of Bodie and Mammoth City, about 11 miles southwest of the former and 45 miles north of the latter. Rich placer diggings were found there, but were soon worked out. In those days there were at least 1,000 people in and around Mono. Aurora, Nevada — 18 miles northwest — had also become quite an enterprising mining center. Four years later than the settlement of Dogtown, the first white man, as a permament settler, had built his house on the site of Old Fort Independence in Owen's valley. But of this, more hereafter.

DR. RANDALL'S SEARCH.

In 1861, amid lively times, quite a stir was created in this mining camp of Mono by the arrival of a stranger, who was soon learned to be a Dr. Randall, from San Francisco. It then became known that he wished to employ some one to go out prospecting with him in a certain direction. He engaged a man, and with him went out and located a quarter section of land on "Pumice flat," some 37 miles south of Mono, and about 8 miles north of where Mammoth City now stands. He was pronounced by the miners a fool for making such a location, and he gave no reason for it at that time, except that he liked the looks of the country. This is the place that afterwards became known as

GID WHITEMAN'S CAMP

The following spring Dr. Randall returned, bringing a man with him. He then employed as fellow prospector and fore-man Gid Whiteman,[15] and a force of 11 other men. From that date Gid became the head man in this search, and lives and searches around there even now. This time he brought a map with him, and, using it as a guide, he and these men thoroughly prospected his 160 acres. He found some reddish lava, or cement, as it was then and has ever since been com-monly called through all that region. Believing he had found the outcroppings of the valuable ledge from which his gold-bearing cement came, he placed in possession the man he had brought with him. He immediately returned to Mono and showed, for the first time, some fine rock—a reddish cement. It was undoubtedly rich in gold. Many reliable people who saw it declare that at least half of it—Mark Twain says two-thirds—was composed of large, pure flakes of the precious metal. It is now supposed that this was a piece of the ore the dying man had given him, but, as he told nothing of this, the impression was that he had found it in his late prospecting. A tremendous excitement was the result. Prospectors poured out of Mono and Aurora day and night, watching and dodging each other, dogging each other's steps as they sought eagerly—nay, frenziedly—for such cement. Never was there a greater furor around any mining camp. This was in June, 1862. Among others drawn, some time afterwards, into this mad rush after the precious red cement was no less a personage than Mark Twain, as he admits in his inimitable sketches of such early mining excitements and their disasters in his "Roughing It." What reader has not been convulsed with laughter and thoroughly well entertained by his humorous recital of the untoward incidents of a secret expedition leaving Esmeralda after midnight, under Whiteman's guidance, in search of this "wonderful cement," which, he asserts, was "fully two-thirds gold." (Chapter XXXVII.) This was several years after Gid began the search. Recall the sudden unhorsing of Twain in the dark, near "the last cabin," his relief, and how next day "the rest of the population filed over 'the divide' in a long procession," their discovery stopping Whiteman's hunt, as usual. Recall all this and the subsequent visit to Mono lake and enjoy a hearty laugh again, but remember the real facts of the original "find" were not as well known then as now. Friend Mark was giving us humor rather than history.

HUNT FOR CEMENT MORE THAN HALF GOLD.

Through the whole summer of '62 hundreds of prospectors hunted industriously for the red cement, but they found nothing like it. At the same time all sorts of stories and talk were in circulation. Some men abused the Doctor, said he was a humbug; others defended him. Some said the cement he showed was manufactured in Aurora; others declared it came from South America, but most of them thought it must be genuine, and that the Doctor was an earnest believer in the existence of such a ledge, for he was spending money freely. It is certainly true that a large piece of such cement as is

described here was kept for a long time on exhibition in
Chapin's, afterwards Frank Schoenmaker's saloon, in Aurora,
and a piece of it was preserved carefully for years in San
Francisco and is supposed to be in some one's possession in
this city now. A very similar piece of gold-bearing cement,
or lava, which came from the goldfields of Australia, can be
seen in the cabinet of the California Academy of Natural
Sciences.[16] Large lumps of pure laminated gold—forming,
with the reddish brown mass of decomposed quartz in which
it was imbedded, very much the same looking ore that this
red cement was reputed to be—was dug, to the amount of
thousands of dollars, from A. O. Bell's remarkable "Life
Preserver mine," four miles from Auburn, Placer county, in
the spring of 1877. Such was the genuine faith in the finding
and present existence of large amounts of just such gold ore
somewhere near the locality indicated that hundreds of men
in Mono, Inyo, Fresno, Tulare and elsewhere in California
BELIEVE THAT IT WILL YET BE FOUND.

So strong has been and is this belief, that from 1862 to
the present time, not a year has passed that from one to
twenty parties have not spent much of the summer hunting
for this cement. The search for it was not always by any
means amusing. Many really tragic events have been connected with it. At least seven men are known to have been
murdered by the Indians while looking for it, during the
"Owen's River war." How many more lives it cost can never
be fully known, because of the great secrecy with which the
pursuit of this object has always been made, for reasons that
will become obvious as the events of this sketch are further
unfolded.

But good evidence exists, in the possession of a few persons
only, that such red cement, quite as rich in gold as that
described, *was* found in 1862, during Dr. Randall's second search, though not by him or Gid Whiteman, and
possibly not on his 160 acres, but by two men of their
party, who, in a most remarkable manner, concealed the fact,
and used the discovery for themselves, securing from the
ledge thousands of dollars. According to well authenticated
statements existing, this is true, and all that prevented their
taking out more was the events of the Owen's River war and

the death of the finders before they could avail themselves again of their valuable but dangerous knowledge.

All this, however, must be reserved for a future chapter on some of the later experiences of the "Cement Hunters," and the undoubted and solid practical results therefrom.

The recital of the testimony referred to, in its full details— which will be recorded—also gives very satisfactorily the reason why later prospectors for this lost mine have as yet failed to relocate it.

Episodes & Later Incidents of Cement Hunting

Early in 1861—the very year in the summer of which Dr. Randall made his first search for the "Cement mine," and located "Gid Whiteman's Camp;" and late in the fall of which the Owen's River war began—a stranger came to the town of Mono. His subsequent movements, together with the mysterious and horrible death of another stranger who prospected with him, have ever since been associated by many with the hunt for this same gold-bearing red cement. Bodie papers have lately announced the finding of the head of Hume, a murdered man, in a well petrified condition. This head, now unearthed by mining processes, eighteen years after death, belonged to the man who lost his life as above alluded to. It has a remarkable history, the main facts of which we will here give, as received from good authority in Mono county last summer.

The newcomer to Mono in the spring of '61 gave his name as Farnsworth. He offered to furnish an outfit and give a share in a rich claim he professed to have found. It is believed he succeeded in getting a man to go with him. Some say they were seen going out of the town together, having a small outfit on a horse, both walking. What became of the second man none know.

⊷❧ DEADMAN CREEK ❧⊶

"A searching party of thirty men went out with Indians for guides, to track him. They easily traced him to
the north branch of Owen's river, now called Deadman's
creek, formerly Hume's crossing, or Murderer's creek.
There the Indians found blood, hair and ax marks on a
log. . . . Farnsworth's tracks were traced up and down
the creek to where a man's head was found in the water
covered with rocks. Near by they found the body covered in the same manner."

DEADMAN CREEK

FARNSWORTH'S RICH CLAIM AND HUME'S MURDER.

Farnsworth came back alone, and tried to persuade parties to go with him to a rich carting claim somewhere on Owen's river, generally stating that it was about a day's travel from Mono. No one then made any arrangements with him, as all were busy. He disappeared and afterwards came again with a man said to have come from San Francisco, a drayman, whose name was Hume. The latter had a dray and a large horse. They spent some days making preparations for prospecting. By this time the Indian troubles were brewing. They got a small outfit and went—no one knew or cared where. Shortly after Farnsworth came back without his hat, and on the horse. He told how they had been attacked by Indians, and the other man was killed. He had a bullet hole through his boot-top, and holes in his clothes that were apparently cut with a pocket knife. He then related all the incidents of an attack near the head of Owen's river.

As he did not tell a very straight story when cross-questioned, the people began to suspect him. The crowds commenced talking about the other man, wondering what had become of him. Like most mining camps in early days, Mono had no officers of the law, but the miners, being a law to themselves, in their usual way appointed men to take Farnsworth in charge till they could make search for his missing companion. A searching party of thirty men went out with Indians for guides, to track him. They easily traced him to the north branch of Owen's river, now called "Deadman's creek,"(17) formerly Hume's crossing, or Murderer's creek. There the Indians found blood, hair and ax marks on a log. No Indian tracks appeared anywhere around. Farnsworth's tracks were traced up and down the creek to where a man's head was found in the water covered with rocks. Near by they found the body covered in the same manner. As soon as the head was found a man was sent back to Mono to have Farnsworth's guard increased till they could return. They buried the body at the root of a pine tree, and carried the head up to Mono the next day. This head was preserved in whisky, and afterward identified by Hume's sister, who came up from San Francisco for that purpose.

Unfortunately for the ends of justice Farnsworth escaped that night, while the guard had become drowsy and relaxed their watchfulness. He was traced as far as Honey lake, in what is now called Lassen county, but was not caught, nor have any of the Monoites ever seen or heard of him since. Then the question was asked "What could have been the cause of this bloody deed?" Some surmised that Farnsworth may have tolled Hume out to murder him for his money. But the latter is believed to have had very little with him. Hence it is generally considered more reasonable to conclude that Hume was murdered to prevent his recalling the location of the rich claim, which is thought by many to be identical with the wonderful cement mine which Dr. Randall was looking for.

UNTUTORED INDIAN TASTE VERSUS THE VITIATED ONE OF THE WHITE MAN.

One more incident we must record about the foul deed, not to horrify our readers, but to adhere to facts. You have already been told how the Indians feast on slimy maggots and hairy caterpillars—ideas well calculated to turn a civilized stomach. We can well imagine that some of our fair readers of such facts said with a shudder, "Ugh! it makes me sick. I don't like to think of such things. Those hideous Indians?" Yet it is simply true, as is also what we shall now tell. We forewarn those who do not wish to read what is coming to "skip it."

You have been informed that Hume's head was preserved in whisky for a time for identification. It was placed in a keg and well covered with this popular antiseptic fluid. An old toper, far gone in the lowest stages of intemperance, where craving thirst for the stimulus of alcohol rules supreme, regardless of taste, found out the liquor was there—going to waste, as he thought. So he secretly drank nearly all the whisky from the keg before he was detected and stopped. Just think of it! Was this not cannibalism? Which is most to be condemned—which most shocks us: the rude, unculti- vated taste of the untutored savage, or the depraved thirst of the degenerate, though once civilized white man? Were any of us so placed as to be forced to try either one or the other, which would we choose: this white man's dram, or the Indian's worm or caterpillar feast? Why, we would try the Indian diet, of course, even if we took an emetic immediately afterwards—that is, should we need one. The keg was finally buried near Mono, with the head in it. Afterwards it was hydraulicked out, and reburied in what was supposed to be a safe, last resting place. But here it comes to light again in 1879! For it is certainly this head that has recently been mined out, now well preserved by petrefaction. Reader, pardon the question: When will this exhuming process end?

OTHER LIVES LOST IN THE SEARCH.

Sometime in the fall of 1862 a teamster was driving near Mono lake, on his way to Aurora. At considerable distance from him he saw an odd-looking object moving along the shore, close to the water's edge. He could not tell at first whether it was man or beast, as it was on all fours and moved slowly. It was coming towards him, and he drove out of his way to meet it. It proved to be a man—the only one left from a party of three prospectors that had been attacked by the Indians while camped on a bank of the creek above the falls to the northwest of where Mammoth City now stands. The remaining two, he was sure, were killed. He was wounded, but jumped into the creek, and was carried over the falls. Though very badly injured, both by the bullet and by the rocks in his descent, he managed to hide by holding on till dark to some bushes on the edge of a small lake where he found himself. It was this, or instant death, if he showed himself. At last, nearly frozen, he managed to climb out, and, with the greatest difficulty and pain, made his way, chiefly by crawling, to the point where he was found, some thirty-five miles from the place of the attack. He was taken to Aurora, recovered, and was afterwards a butcher there. He and his comrades were hunting for the cement when attacked. The trip satisfied him, and he never searched for it again. Indeed, this and similar events greatly checked the search, until the close of the Owen's River war. No prospectors then, or for some years afterwards, ever hunted this cement that one or more Indians were not found dogging their steps, sometimes seeming almost to rise out of the ground.

MANY A SECRET EXPEDITION

For this purpose has been made, of which no record has been kept. It is well known that different parties who hunted this cement never returned; that men lost their partners as mysteriously as Bodey disappeared, whose name, with a slight change, is preserved in the fine mining town of Bodie to-day, and whose body has lately been discovered and decently buried there. But those of the cement hunters that suddenly disappeared were usually supposed to have been murdered by the hostile Indians. As already mentioned, seven men are known to have been killed in this connection, and there were, no doubt, others. Still, this perservering search was never entirely abandoned, even while the war was raging and surrounded the work with so many dangers.

Certain it is, that the Indians drove out nearly every party so long as Joaquin Jim lived. He made his headquarters in Long valley and vicinity. Since 1867, there has been comparative safety from Indians, though they are still saucy, boasting even now that they whipped the white men in their war. According to all reports, since gathered, some of them have long known and valued this ledge of free gold which the white men have so industriously sought. Those of them who have ever alluded to it, always represent it as near a lake. Many a time since the war prospecting parties have tried, by all sorts of inducements, to get them to show such gold-bearing ledge. But the only means by which they have generally been induced to serve as guides has been by talking to them of silver, which they call *monie*. Gold they call *pijyah monie*. These Indians have a belief that gold is what brought white men here, and when the gold of our mountains is exhausted, white men will leave. A proposition correct in the main, but admit it to the fullest extent, and there is no doubt gold enough in our mountains yet to keep white men searching it for thousands of years to come. But the red man, as he is gradually passing away, cannot see this. Hence his unwillingness to help us find gold.

Another trouble with the Indians is, they never read Virgil —not even in a translation. Consequently, they don't know anything about that remarkable exclamation, so wonderfully suggestive and full of truth: *"Auri sacra fames, quid non mortalia pectora coges;*" or in plain English, "Accursed thirst for gold, what will you not force mortals to do!"

That these Indians do or did formerly know where some valuable gold mine is, the following well authenticated fact proves:

A cattle dealer by the name of Jones, from Santa Barbara, once came to the ranch of Mr. Wilson, [18] the pioneer settler of Owen's valley. In conversation he asked him where the Long Valley Indians got their gold. He showed him about $300 in grain gold from the size of wheat to a good sized bean, which these Indians had paid him. But we are somewhat anticipating, and will now relate the circumstances of one of the best sustained searches for this wonderful cement mine, after peace was made with the Indians.

KIRKPATRICK'S EXPEDITIONS.

In 1862, after Mr. Wilson's home had been abandoned by him, and turned by the soldiers into Camp Independence, and the war was at its hight, he had moved to a more quiet region and located a ranch near the town of Mono. His house, being on a much traveled road, half a mile from town, became a general stopping place for prospectors. From them he heard all the talk about the cement hunting. He paid little attention to the affair, however, believing it was like many other mining stories, all "breyfogle," as the miners say, from a German of that name, who became as notorious for the absence of truth in his mining tales as Baron Munchausen was among travelers. But the following occurrences entirely changed his view of the matter.

In 1865 a well equipped company that had sold out a fine mine in Idaho called "The Poor Man's Ledge" came to this place from Sonora, Tuolumne county, to make a thorough search for this now noted cement ledge.

George S. Kirkpatrick had charge of this expedition, and "Si" Colt, nephew of Colonel Samuel Colt, of pistol celebrity, was with him. They wanted a guide. About the following colloquy occurred between Kirkpatrick and Wilson:

K.—You have had some fine diggings here, haven't you?
W.—Yes; but most of them are worked out.

K.—Didn't you have considerable excitement about a rich cement ore some time ago? W.—Yes.

K.—Do you know where it was said to be? W.—Yes. On the headwaters of Owen's river.

K.—Did you know a man named Van Horn? W.—I did.

K.—Do you know where Gid Whiteman's camp is? W.—Yes; very well.

K.—Will you go and show us the place? It is all we want to know. W.—No; for I believe the whole thing is a humbug. It has been sought for so much without a trace of such gold.

K.—But will you go if we make you know it is not a humbug? W.—Yes. If you really convince me it is not a humbug I am willing to show you where Whiteman's camp is.

They then brought out a carefully written description of the locality, told him all about Van Horn's death, and how he had really found the rich, gold bearing cement in 1862, when he was one of the twelve men employed by Dr. Randall in his long and faithful search for the ledge. But these facts and subsequent events we must reserve for the next chapter.

SITE OF MAMMOTH CITY
LAKE DISTRICT,
AT THE BASE
OF MINERAL HILL

"Gid Whiteman, among others, acknowledges that he has been out every year since '62 there, and in neighboring localities, hunting for this rich cement. He was in Mammoth City every once in awhile last summer. . . . The discoveries in 1877 of the Alpha, Mammoth and other promising ledges of the Lake district, as well as the Laurel district of Mono county, come directly or indirectly from this long search for the celebrated cement mine."

SITE OF MAMMOTH CITY

The Cement Hunting
·→❧· Continued ·❧←·

We shall here give in detail the incidents related by Kirkpatrick and his friends, which, with the subsequent search,
fully convinced Mr. Wilson, in spite of his fixed skepticism
on the matter, that two of the men in Dr. Randall's party *did*
find this valuable red cement ledge in 1862, and profited by
it for a time. These things were told as secrets then, and
comparatively few know all the facts now. But the ban of
secrecy is at last removed, and they may as well be made
known to all who choose to read them. Now for the convincing statement.

VAN HORN'S STORY.

An invalid named Van Horn was going by boat from Sacramento to San Francisco, and his friends—Carpenter and wife,
of Aurora—were with him. His physicians had told him that
he might die at any time. On the trip he was taken suddenly
ill, and called Carpenter to him. He said he felt that he was
sinking, and, perhaps, would not live through the trip. He
thanked Carpenter for his great kindness, acknowledging that
he had acted like a brother towards him, although he had no
reason to think the sick man could make any returns for his
careful nursing. Yet he could and would repay him. He told
of his connection with Dr. Randall's expedition in search of
the cement. He was cook on that trip. One day the man that
the Doctor had brought up with him, and who is said to have

been the friend of the man who first found the cement, came in, very much excited, when Van Horn was alone. He had always declared the Doctor was looking in the wrong place. He was a German. Finding Van Horn by himself, he said to him, in confidence, "I find dot d—— ting, and I get noting but mine wages." At the same time he pulled out of his pocket a lump of the cement, rich with free gold. Van Horn, full of excitement, told him, "Hush, hide it; we'll fix all that. I get nothing but my wages, either." They mutually planned to get discharged next day, complaining to the Doctor that they were dissatisfied and did not see "the use of hunting gold in a country where the rocks swim and wood sinks." This referred to the vast amount of pumice scattered all over this volcanic region, and to the mountain mahogany (*Cercocarpus ledifolius*), a very solid, heavy wood. They packed what few things they had on Van Horn's horse and started out towards Aurora. But, as soon as they were out of sight of camp, they turned to one side and went in the opposite direction, to the place where the Dutchman had found the cement. They dug down only two feet and became terribly excited, because they saw it was more than half gold. They filled a sack with the precious ore; then they quit digging, that they might go and get a good outfit, for they had food for only a few days. They buried their pick, an ax, and other things they would not need till they came back. They covered the exposed part of the cement well with pumice, took a bush and swept it over to deface all marks and make it look as much as possible like the surrounding surface. Then, as they left, one of them walked backward and swept out their tracks. Carefully avoiding Mono and Aurora on their route, they hurried on to a point on Walker's river known as "The Elbow." There they crushed and panned out their sackful of ore,

GETTING IN ALL $30,000.

This amount is by no means incredible, if the ore was over half gold. For, the entire weight of the sackful would not have been much, if any more, than 175 pounds, and the horse and two men could easily manage that, and what little else they had to carry. They then went to Virginia, got a winter outfit, took in a partner and started back. Meanwhile, about a week after these men left, Gid Whiteman and his party quit work for Dr. Randall, and the latter disappeared from the scene.

Van Horn and his two companions, on their return to the ground where they had buried their tools, arrived at two P.M. They at once began their preparations for building, cut down some suitable trees and selected the site for their cabin, because they proposed to winter there. The next morning, while at breakfast, the three men were surrounded by Joaquin Jim and thirty of his warriors. The Indians surprised them, took everything they had but an old horse, destroyed the camp and ordered them out; told them they should not work there; said they had known that place a long time; that what the white men were looking for there was medicine, and the Indians wanted it.

Glad to escape with their lives, the men left, and at a safe distance stopped and made each other a solemn promise to tell nothing of it till they met again, when the Indian war was over. As it continued nearly three years longer, they were scattered and never met again. Two were reported to have gone off to the greater war East with the "California Hundred," as Van Horn believed they had done. The third was the sickly man, Van Horn, who told Carpenter this story, and gave him certain directions for finding the place.

DOUBTS REMOVED AND THE SEARCH RENEWED.

Convinced by this array of facts and minute details, and by the confidence in them on the part of men of capital who had lately been successful in their Idaho venture, Mr. Wilson consented to go with them, as they requested. Under his guidance, and with some Indians he induced to go with him, as his friends—for he had always retained their goodwill by his personal kindness to them—they started for "Gid Whiteman's camp." They spent about six weeks hunting over all the adjoining country. In several places they found a red cement similar to the gold-bearing ledge described, but no gold in it. They scoured the country well, yet were not satisfied. So they got a new outfit and returned with fresh

Indian guides. Even some of the old "cement hunters" of
Dr. Randall's party went with them. An Indian chief told
them he had seen the two men bury their tools and sweep
the ground, and could show them where it was. He started
out, and led them to a destroyed cabin. Then he complained
that he had been badly treated, and left them to hunt the
buried treasure for themselves, as these sly red men so often
did on such excursions. Again, an Indian chief showed one
of these hunters what he said was a destroyed camp. Here
were found a broken shovel, ax, cracker boxes, barrels, a
candle box, some forks, a miner's hornspoon for prospecting,
a mashed coffeepot, and near by a sheath knife. The letter
V was found on the knife handle and ax helve, supposed to
have stood for Van Horn. In fact, the knife was afterward
recognized as Van Horn's by those who had known him well.
On one of these expeditions an Indian volunteered to show
a camp of white men that they had killed during the war.
He did so. It was within a few rods of the camp above
described. Here was found a brush camp and a man's bones,
and other marks tending to confirm previous statements.
Another skeleton has since been shown near there by another
Indian. From the finding of these two skeletons, coupled
with other circumstances, it is conjectured that the two companions of Van Horn, who were reported to have gone with
the California Hundred, gave Van the slip, returned to the
camp, and were there murdered by the Indians. This is certainly quite probable. There are really but few persons living
who know the exact directions, as shown by Kirkpatrick and
his party. Though some of these think they know

THE LOCALITY OF THE DESTROYED CAMPS,

None, so far as is generally known, have ever found the gold
bearing cement there since the days of Van Horn and his two
comrades, and perhaps Farnsworth. Can we wonder at the
difficulty of finding it, if it really does exist, when we remember how thoroughly the place was concealed by the two
men who took out the $30,000, so as to make it look just like
the rest of a large area covered for many square miles with
a grayish white pumice, from the size of coarse sand to a
hen's egg? Furthermore, is it surprising that from statements

made that are deemed reliable, many prospectors have become imbued with unswerving faith in the existence near the surface of such a ledge, somewhere along the head waters of Owen's river? Gid Whiteman, among others, acknowledges that he has been out every year since '62 there, and in neighboring localities, hunting for this rich cement. He was in Mammoth City every once in awhile last summer, but was spending most of his time in Prescott district,[19] where he has some very good ledges of gold and silver ore, to the discovery of which he was led by his persevering search for the noted red cement, with which his name will be handed down, while mining and its history exist. Mr. Wilson, since his conversion in 1865 to the belief in its existence, has been among the most persistent, intelligent and best posted "cement hunters."

The same Indian who showed him the bones near the destroyed camp also guided him to a place where some one had split an old log into puncheons, made a floor of them, put up side pieces, and had kept a red, rusty looking quartz there. Here in the cracks he found crumbs of this rock, rich in free gold.

AS A PRACTICAL RESULT

Of his search, which all such facts stimulated, he now owns at least ten fine ledges of silver bearing quartz in the Minaret district, Fresno county, but a few miles west and southwest, across the Summit ridge, from the now classic scenes of the exploits and sufferings of the "cement hunters." In 1865 his Indian friend, Shohock, who guided him in Kirkpatrick's second and third expeditions, led him to one of the ledges he now owns and told him, "*Tuppee* (quartz) along here." In his honor Mr. Wilson called these ten-wide ledges "The Shohock Consolidated." So have the discoveries in 1877 of the Alpha, Mammoth and other promising ledges of the Lake district, as well as the Laurel district of Mono county, come directly or indirectly from this long search for the celebrated cement mine. We can safely give as the practical result of this search, which to many may seem like madness, the discovery, in less than three years past, of more than two hundred valuable mineral ledges, chiefly silver, but nearly all with some gold, in the adjoining districts of Mono and Fresno counties, on the headwaters of Owen's river and the San Joaquin, in that grand and wonderful, almost enchanting, mountain region of California.

IN THESE DAYS OF SPIRITUALISTIC TENDENCIES,

Noted mediums have been consulted about this lost mine. Those who believe in them contend that their communications about the rich ledge agree with the less ethereal and more practical records—as is usually the case. They are said to assert that "It does exist; that a cloud of Indian spirits are always guarding it; that its rediscovery is reserved for a poor man, but is death to the finder." Now, anybody may believe all this that wants to, but in the opinion of the writer a good combination of pluck, energy, perseverance, muscle, drilling, giant powder, blasting, with some knowledge of mineral formations, is far more reliable to solve all such questions than the whole mass of imaginary spiritualistic dictums that can be stirred up. Many a poor man has risked his life and will take the chances on it hereafter to find that and similar wealth-giving lost gold mines. Now that Indian flesh and blood don't interfere, what need he care for Indian spirits? Indians have more reason to fear the white man's "spirits," if they only knew it.

·❧ A Queer Episode ❧·
Who Was He?

Late in May, 1869, two men arrived in Stockton, as fellow passengers on the overland train. One of them was middle aged. His hair and full beard were slightly tinged with gray. He was above the medium hight and of powerful physique. His companion appeared to be at least ten years his senior, and of slighter build. They represented that they were from the neighborhood of Salt Lake City, though they had both been in California in its earlier mining days. We shall call them Kent and McDougal, though they then gave different names. They remained in Stockton only long enough to buy a strong, four-horse wagon, four good-sized, half-breed mustangs, well broke to harness and saddle, some mining and other tools, a considerable supply of provisions, a dozen strong packsaddles, and altogether a very good outfit for several months of camp life. They were soon on their way by the usual road to Knight's Ferry,[20] and thence along the foothill route to Snelling and by way of Union Postoffice to Jones' store, on the south bank of the San Joaquin. They arrived there about the middle of June. This was quite a noted supply point, kept for many years just below Converse's ferry, by J. R. Jones, a well known pioneer of Fresno county, now gone to his long home. It is located just where the second largest river of California leaves the foothills of the western slope of the Sierra Nevada, two miles below Millerton, and enters the broad plains of the San Joaquin valley,

RAINBOW FALLS MIDDLE FORK OF THE SAN JOAQUIN RIVER

"Their first care, however, was to put up a rough cabin. . . . On the low, grassy meadows along the river there was plenty of feed for their animals. . . . These matters provided for, they cautiously crossed the great divide between them and the headwaters of Owen's river, and, after a little careful searching, Kent finally identified certain landmarks, which he had retained in his memory, and found a reddish ledge, rich in free gold, though not very wide. On this they worked off and on for about two months, spending part of their time hunting deer and catching mountain trout below the falls, ten miles down the river from their camp."

RAINBOW FALLS

which, including Tulare valley, extends southeast of that
point nearly to Fort Tejon and Tehachapi valley, about 150
miles in a direct line, and northwest to Stockton over 100
miles—varying in width, between the Coast range and Sierras,
from 40 to 60 miles. Somewhat replenishing their supplies at
Jones' store, they at once entered the mountains, going northeast by road to

CRANE VALLEY,

Entering its southern end where now stands the ruins of
McCullough's sawmill, which then, and for so many years
before, supplied the surrounding country with lumber for
many miles. This fine mountain valley, noted then as now
for its productive grain and hay ranches, is about five miles
long from north to south and half a mile wide, its elevation
above sea level being about 4,000 feet. Through it runs what
is called locally, and even on some of our maps, the "North
Fork of the San Joaquin." Though this is a good sized mountain stream, thirty feet wide in places, and runs all summer
even in our dryest season, it is really, like the Chiquito
Joaquin, only one of the numerous smaller and more westerly
branches of the San Joaquin, while the true North Fork is at
least thirty miles farther northeast.

Here the two men made arrangements to leave the wagon
at a ranch for several months, bought and hired enough pack
animals to carry their supplies over the rough mountain trail
through and beyond Beasore meadows, and secured two
Indian guides to show them the route to the headwaters of
the San Joaquin. The man we call Kent told the ranchers he
met that he was acquainted with the country in Mono and
Inyo counties, had some friends over there he was going to
see, and that all he wanted the guides for was to lead him
to the pass near Pumice or Mammoth mountain into Long
valley, on Upper Owen's river, for, beyond the pass, he knew
the trails. From that point he promised to send back the
hired donkeys by the Indians, and for this the wagon and
harness left were considered good security. Now, all beyond
the matters here stated that was known by the settlers along
the line of travel, then or ever since, about these men and
their objects, was that the Indians returned in due time with

the hired animals, reported that the white men had gone over the mountains, and were well pleased with the trinkets and blankets given them by these men for their services; also, that the men returned in the fall, and made similar successive trips every year till the summer of '77, since which time no one in that wild and sparsely settled region has seen or heard of either of them.

A MYSTERY SOLVED.

But a late explanation comes from an unexpected source, and is as follows, though no real names have been divulged:

Late in the fall of '77, a man somewhat advanced in years fell senseless upon one of the streets of San Francisco, from a stroke of paralysis. He was removed to one of the city's hospitals in a prostrate condition, though he lingered for some time, and rallied sufficiently to talk freely to those around him. Previous to his death he called for a father confessor, and among other things, told him in substance what will now be related: Previous to 1869 he had been a prospector and miner in Utah and adjoining territories. Near Salt Lake City he became acquainted with a man known there as Kent, also a miner, but who came only occasionally in summer to Utah, spending much of his time farther East. He did not know exactly in what locality. As their intimacy grew, Kent seemed to take a fancy to him, and one day told him, in strict confidence, that he knew of a good thing in California, where he once mined. Finally, in speaking of this matter, Kent said to him that he had some trouble in California several years before, but that so soon as the railroad was finished through to Sacramento, he wished to return, and that if he would go with him, pledge himself to profound secrecy about the work he proposed and its results, and remain faithful to him, he would guarantee the punctual payment, whenever wanted, of a salary of $1,500 a year. This compact between them was made, and they were the two men who came to Stockton in May, 1869, as above described, and made the trip to the upper San Joaquin then and in successive summers. But now comes

THE STARTLING PART OF HIS REVELATION.

When the Indian guides left them, just beyond the ford of

the main San Joaquin, but a mile or so west of Mammoth mountain, they did *not* cross the summit ridge immediately into Mono county. On the contrary, Kent then informed him, that now was the time for profound secrecy and the greatest care; that they were near what he thought was one of the richest gold ledges of California, and it was to find this again and work it that they had come to this mountain wilderness; that he had found it in '61, but the Indian war and later circumstances made it impossible for him to get the benefit of his knowledge; yet, he thought he could readily find it again. Their first care, however, was to put up a rough cabin, to protect their supplies, and to make themselves passably comfortable for the summer. For this purpose, they selected the most secluded spot anywhere near. They found it by following immediately along the base of the main ridge east of the river, turning north from the trail. Following up the eastern bank of a small stream, they crossed a second one just below where it rushed down the mountainside in a pretty cascade over an odd little grotto in the solid rock, a kind of reddish lava. Beyond this and near the base of a small but beautiful lake of blue water

NESTLING IN CHARMING SECLUSION

between the main ridge and one of its spurs, they built a roomy cabin. On the low, grassy meadows along the river there was plenty of feed for their animals, and the latter showed no disposition to leave them in their isolated position.

These matters provided for, they cautiously crossed the great divide between them and the headwaters of Owen's river, and, after a little careful searching, Kent finally identified certain landmarks, which he had retained in his memory, and found a reddish ledge, rich in free gold, though not very wide. On this they worked off and on for about two months, spending part of their time hunting deer and catching mountain trout below the falls, ten miles down the river from their camp. In passing we shall mention that the mountain trout of California differ from those farther east chiefly in having black, instead of red specks. In the opinion of the dying man—who, though known by a different name in the hospital, was no doubt identical with the McDougal mentioned above

—bullion to the amount of at least $40,000 was taken out by them that summer. This they ran into small bars of about $2,000 each, and distributing it among their different packages, when they moved out, they easily avoided exposing any of the treasure they had thus secured. When they were ready to return to Crane valley, toward the end of September, they carefully covered up with pumice the marks of work they had done on the ledge, and left their few mining tools in a secure place. Arranging with the farmers of Crane valley to take care of the pack animals for eight months, and assuring them they would return the following June, they soon made their way back to Stockton, disposed of their wagon and team, and hastened to San Francisco. Here Kent had his bullion coined at the Mint, paid McDougal the balance of the $1,500 due him for the first twelve months, and transferred the bulk of his coin

TO CHICAGO.

They then returned in company to Utah, where McDougal spent the winter, while Kent—promising positively to meet him again the following May, went direct to Chicago—where else McDougal did not then know. Kent was described by his friend as a rather peculiar man, very restless, watchful and silent rather than talkative, though invariably kind and considerate in their intercourse. There seemed always to be an anxious care resting on his mind.

McDougal then told how this secret work was continued by the two year after year, until the summer of '77; how the deep seclusion of their summer camp was never during that time invaded by white men or Indians, because the few who ever passed near it went either along the dim trail south of them or along the main San Joaquin west of them. He positively declared that they took from the ledge each year from $25,000 to $50,000 in gold, he always faithfully keeping his pledge of secrecy, and Kent increasing his pay to $2,000 for each summer's work. To be sure of this amount, he said, was all he cared for and his wants were few. He was impressed with the fact that Kent never went anywhere east of the great divide between Fresno and Mono counties, except to their ledge and back. Whenever any additional supplies were needed, McDougal, well armed, went by Pumice Gap trail to Benton, forty miles east, or Bishop creek, fifty miles southeast, and got what was wanted.

THE PLEASANT MONOTONY OF THIS WORK,

And its pleasanter results, were varied during only two years. By invitation, he went East with his friend in the fall of '74, visited many Eastern cities, and spent several weeks at Mr. Kent's home. This he found to be on one of the most productive, best stocked and most beautifully improved farms in a western state, within 150 miles of Chicago. There, with a happy wife and family, the man we here call Kent lived the active, independent life of the wealthy farmer; a highly respected and trusted citizen in the community in which he was known. But he persistently avoided notoriety, shunned, rather than sought society and promiscuous crowds, while the keen observation of his true, tried and devoted friend did not fail to detect that same restless caution, and a mental anxiety at times, as if a canker was gnawing at a happiness which otherwise would have seemed complete. Yet, while McDougal felt satisfied that a dark shadow rested on some part of his friend's past life, the fact that he had seen him in his delightful home, a beloved husband and father, honored by a large circle of relatives and friends, made him resolve the more firmly to preserve his fidelity towards him in every way. Does some one say "Such fidelity is rare?" True—but is it impossible? Their summers in the higher Sierras had heretofore passed so quietly, without any disturbance from Indians or other sources, that Mr. Kent concluded to take his family with him to California the following May, that they might enjoy the rough but healthful mountain life, in their wild camp through the summer. He merely charged them most strictly to say nothing to any they met about the object of their visit and its work, merely having it understood that they lived in the country near Chicago and came to California for a few months to see its curious features and to visit friends.

A WILD MOUNTAIN HOME.

By the middle of June, without accident, the party reached the cabin in its secluded nook. Two plain log shed rooms were added to shelter the family, consisting of a son—a fine, manly fellow about eighteen years old—a beautiful daughter of sixteen, and three younger children, a girl and two boys. The young people especially, always in the bloom of health, were delighted with their curious mountain home and its grand surroundings for the two months and more they spent there. It was so odd to be so completely out of the world, and to see nobody all that time but their own household. It made them think, many a time, of the "Swiss Family Robinson." Then everything around them was so different from anything they had ever seen or dreamed or read of, unless it was some of the wilder parts of the Alps of Switzerland.

A favorite place of resort for the rustic sports of the children and an occasional family picnic was along the pretty babbling brook near the very picturesque cascade and grotto already mentioned. These together certainly form one of the choicest gems of nature in one of her own most charming settings. The children proposed to call this Lily grotto, after their oldest sister, who bore their mother's name. But she wanted to name it Pumice grotto, because it is at the western base of Pumice or Mammoth mountain, and because the smooth rock over which the water dashes and in which the grotto is worn is of that pale, brickdust colored lava, of which this vast and lofty dome is composed. The matter was finally compromised, so that they were christened respectively

PUMICE CASCADE AND LILY GROTTO,

The hand of Nature's great High Priest furnishing the water for the sprinkling. Let us try for a moment to trace a faint word-picture of this beautiful gem of nature as it appears in July. We stand among the handsome firs, tamaracks, junipers and mountain poplars, or aspens—somewhat scattered here. We are near the limpid rock, about twenty rods from the base of the falls, facing eastward. The precipitous but green and sod-covered mountain side rises abruptly at an angle of over 70 degrees towards the top of Pumice dome, which towers at least 4,000 feet above you. About 80 or 100 feet up is a terrace gradually sloping back some distance and thickly overgrown with cone-bearing evergreens and shrubby undergrowth. Through a shallow notch, at least 75 feet above you, and thickly shaded with rich, green bushes, the crystal waters forming this perennial cascade burst upon your sight. There the stream is only four or five feet wide, though before it reaches the bottom it spreads, fan-like, to fully 20 feet. In its descent of 75 feet or more, its wild water rushes swiftly in ribbons of white foam and spray over the smooth worn, gently curved lava, here gray as granite, and even darker, from the constant wetting it gets. To right and left of the cascade the mountain slopes are beautified by rich mosses, green shrubs and flowering plants in considerable variety. A few feet above the seething pool, at the base of the cascade, is the grotto. Let us go nearer, and examine it critically. It is a niche in the solid rock, exactly oval in shape; just as if it were the mold from the casting of half of a huge egg, with the point upward. This very symmetrical concave oval is about five feet high, four feet wide and three feet back to its deepest point. A thin, clear sheet of water rushes over it, like a glass vase inclosing it. In its lower part—somewhat flattened—a small amount of fertile soil has accumulated, and in this are growing several pretty tufts of grass and other plants, thriving and blooming. Among these was one of our several species of very handsome white mountain lilies, others of which are found blooming in midsummer along the streams below the grotto. This was considered another good reason for calling it "Lily grotto" and so it was. Reader, imagine the unique and exquisite beauty of those rich green blades and leaves and pretty flowers, growing and blooming in perfect security in that grotto, chiseled by Nature's sculptor, at the base of a wild cascade, as seen through one of Nature's own crystal vases!

AMONG THE FLOWERS,

Blooming in the middle of July, along the gurgling brook, just below the cascade, were large yellow "evening primroses," orange colored "painted cups," "monkey flowers," light blue lupins, pink geraniums, blue larkspurs, crimson leptosiphons, dandelions, the columbine seen in most of the Sierras, gentians, dwarf thistles with large lilac colored blooms, straw colored Solomon's seals, and large, purple clover, besides the lilies already mentioned, and the pretty flowers of two kinds

of wild onions. Of shrubs, currants, the flowering dogwood
and dwarf willows were in full bloom. This, remember, is
about 8,600 feet above sea level. Here the young folks whiled
away many a happy hour, never being allowed to go further
from the cabin, unless accompanied by their father or Mr.
McDougal and their older brother, each armed with a trusty
rifle, for fear that they might meet with bears or unfriendly
Indians. They were fortunate, however, to escape all such
dangers, though three cinnamon and black bears were killed
during the summer, to say nothing of plenty of fine, fat deer.

Nothing seemed odder to the young people, or amused
them more, than to throw even the largest lumps of pumice
into the streams without its sinking, as they thought all stones
should. Many a time they made the woods ring with their
merry laughter as they saw this pumice bobbing up and down
on the surface like a fishing cork as it rushed over the rapids.
The red and purple gravels of lava and the black, glasslike
obsidian pebbles and small bowlders that abound in the
larger streams also seemed very curious to them.

But why turn aside so long from the more practical pursuit
of gold—the absorbing business of life—to give this attention
to the works of nature and the thoughts and pastimes of
children? Reader, do you wish to know? Well, because it is
best through life that none of its important or instructive
features should be overlooked. To its earnest and too often
harsh realities; to its harmless recreations; to refreshing studies
of the wild, the beautiful and useful in nature and in art; to
a hearty sharing of the innocent enjoyments of childhood;
to these, and to all the most practical and valuable features
of human existence, now and hereafter, should a proper time
be allotted that none of our work may be overdone and none
neglected. In this, we believe, is the truest theory and practice of life. But to return to business. The usual amount of
gold was secured that summer. None of the family but the
eldest son ever accompanied the miners to their rich ledge,
and he helped considerably in their mining operations. By
the first of October, they were safely housed in San Francisco,
and after a few weeks' enjoyment of this

MOST EUROPEAN OF ALL AMERICAN CITIES,
They returned East, McDougal stopping in Utah. In '76

Mr. Kent came West a few weeks later than usual, having visited with his family the grand Centennial Exposition early in June. They secured the usual amount of gold that summer, but Mr. Kent's household was darkened with sorrow the following winter by the loss of his oldest son.

On their return to Crane valley and their cabin in June, 1877, Mr. Kent seemed very much disturbed by the news of the discovery, the preceding February, of the Alpha, with other silver and gold quartz ledges near Pine City, and that Mammoth City was being built up by extensive preparations for milling the ore of Mammoth and other mines. Satisfied that the whole country near him would soon swarm with prospectors, he told McDougal that they must hastily secure as much gold as they could from their mine, destroy, as completely as possible, all traces of their work, and leave. He doubted, he said, whether he would ever come back. In accordance with this plan, they had one day thoroughly filled and covered up with pumice all their last openings in the vein, when suddenly they heard distant voices, and looking through an open space among the trees in the pumice-covered valley where they were, they saw several men with pack-animals coming towards them from the southeast, and evidently on a prospecting trip. The newcomers were at least half a mile distant, but the echo from the surrounding mountain sides had brought the sounds to their ears. They merely had time to hurriedly pack their horses, and started westward at once across the great divide. In their haste, however, they failed to destroy an old puncheon floor and side pieces, where they had at times piled their crude ore before working it out. Could this have been the puncheon floor where the old cement hunter found, between the cracks, crumbs of reddish, rusty looking rock, rich in free gold, as described in chapter vii? The next day they sought to destroy every vestige of their cabin and all traces of their settlement there. They tore it down, piled all the logs and puncheons together, burnt and consumed them as completely as possible, threw the ashes in the lake, which they had always known as

MYSTIC LAKE,

And buried deeply under the pumice the small amount of the charcoal remaining. They served in the same way a small bridge they had made across the stream that leads from Mystic lake to the San Joaquin. They immediately, moved out, and on their way to Madera—for in their later trips, they no longer went as far south as Jones' store—they passed numerous bands of sheep, which in that dry year, were driven through the very region they had just deserted, to what is now called Agnew's meadows, near which the first silver quartz ledge in North Fork district was discovered that summer.

Kent had returned East only about two months before McDougal was stricken down with paralysis, but he had given the latter the strongest pledges, that while he lived communication should be maintained between them, and that he would always see, that Mc's wants were provided for.

According to McDougal's statement, Kent and he must have taken out between $350,000 and $400,000 from the rich gold bearing ledge, the exact location of which they thus succeeded in keeping so profound a secret. Though the probabilities are against their mine being identical with the original cement mine, it was most likely an extension of the same ledge. In this connection, it is food for thought that within the last two months rich ledges bearing free gold have been found in what is now called the Homer district,[21] twenty-two miles southwest of Bodie, a locality but a short distance from the destroyed camps of the cement hunters.

Now, the question arises, who was the man that McDougal had served with a faithful friendship equal to that between David and Jonathan, or Damon and Pythias?

WAS THIS FARNSWORTH?

Further investigation alone can answer this question. To the arbitrament of the immediate or remote future we leave all such matters of the now unknown. In this chapter you have all that is known about it.

But should he prove to be Farnsworth and be still living, and should the shrewd detectives of our day finally ferret him out, would he pay the just penalty of this barbarous murder of Hume? Or would his wealth, his consequent influence, and the sympathy for his family, save him from a felon's doom, as is too often the case with such influential criminals?

·❧ THE MINARETS ❧·

"This camp . . . was perched on a most picturesque
notch . . . on the western slope of the summit ridge of
the Sierras. . . . To westward the view is commanded of
the rugged but heavily timbered valley of the upper San
Joaquin. . . . The bed of the main river . . . is about a
thousand feet below you. . . . Up and beyond the river,
the misty looking, bare, gray hights of the Minarets,
Mount Ritter and Mount Lyell in the distance, varied by
occasional red and purple streaks, and the deep gorges
and snow masses along the base of each, was a feast, of
which the eye never grew weary."

THE MINARETS

·⚬⚞ Late Visit ⚟⚬· to the Destroyed Camps

Last summer, while enjoying for some days a miner's hospitality at Wilson's Camp, or as it is facetiously called by the miners, "Wilson City," the writer listened with much interest to the curious details of this long and mysterious hunt for gold, and accompanying circumstances, as given in what is deemed the most authentic form by Mr. Wilson, who is one of the few known survivors of the early cement hunters. This camp, consisting of a canvas tent, a brush screen protecting it on the south and west, and a small space inclosed by fragments of rock for a campfire—a rough blacksmith's forge standing near by—was perched on a most picturesque notch, surrounded by thick growth of large pines, 9,800 feet above the sea, on the western slope of the summit ridge of the Sierras. Near it is a mountain rill of the purest ice-cold water. To westward the view is commanded of the rugged but heavily timbered valley of the upper San Joaquin, whose extreme sources are only from five to eight miles northwest in direct line, is very grand. The bed of the main river, here a clear, rapid stream, from one to four feet deep and thirty to forty wide in midsummer, is about a thousand feet below you, and scarcely a mile distant. Beyond it you see the silvery waters of Minaret creek dashing wildly over rocky slopes, as it issues from a ravine running southward from the ethereal looking but solid needle-peaks of that name, and joins the main river below your point of observation. Up and beyond the river, the misty looking, bare, gray hights of the Minarets, Mount Ritter and Mount Lyell in the distance, varied by occasional red and purple streaks, and the deep gorges and snow masses along the base of each, was a feast, of which the eye never grew weary. The visit to that camp and its magnificent surroundings, its cordial welcome, its enjoyment, will never be forgotten. There the rough notes were made from which much of this sketch is written, thanks to the kindness of Mr. Wilson, the pioneer settler of Owen's valley. There he, with Messrs. Peck and Elsey, made their home, while the summer's prospecting was going on. At another camp, about a mile by trail, near the top of the ridge, to eastward, at an altitude of about 10,500 feet, and a third or older camp at Pott's meadow[22]—some three miles down the San Joaquin on the trail from Fresno flats[23] to Mammoth City, about 8,500 feet above sea level—Messrs. Potts, Hall, Sotcher and Cernrike had their headquarters, and most kindly shared with me their blankets and fare. These seven men in the Minaret district, with thirty or forty more in the North Fork district, three or four miles northwest, around Agnew's meadows and Norris' store—the future Highland City—were the only inhabitants of that whole region on the upper Joaquin, from three to fifteen miles northwest of Mammoth City. Yet it is no doubt destined some day to become one of the most productive mining regions of California. Not even an Indian was seen there, though there were a few around Mammoth. When the time for leave taking came—as it was learned that we were but three or four miles from the famous destroyed camps of the cement hunters, where were found the skeletons of two white men, who were supposed to have been butchered by Indians seventeen years ago—nothing would do but that Mr. Wilson and the writer must make a short secret expedition to the celebrated

RED CEMENT HILL,

And the mysterious old camps. Thence Mammoth could be easily reached by trail. Mr. Wilson also proposed to try a blast at a certain point, that the volcanic nature of that well-

called "Burnt District" might be seen. So, after breakfast,
devoured with mountain appetites—the mercury early that
morning having stood at 30 deg. Fahrenheit—a few tools
were prepared and the faithful, surefooted sorrel was saddled, that had carried me without ever falling over 5,000
miles during the last twelve months, even to hights of 11,000
and 12,000 feet.

It was a clear, calm summer morning, July 14th. Not a
cloud was seen. The sky was intensely blue; the atmosphere
showing that perfect transparency in which our Pacific coast
excels. As we found our way over the high ridge eastward,
the serene stillness, the perfect silence that reigned, undis-
turbed even by the chirp of a bird, unbroken, except by our
own voices at times, was something wonderfully impressive.

The distinctness of every peak, and ridge, and tree in sight,
as we took in the superb view to east and west from the
summit, was startling. Then down, down by easy descent,
passing on northern slopes some melting beds of snow, from
which initial drops were just starting for Owen's river, we
easily made our way, partly by an old Indian trail, to the edge
of that widely extended

PUMICE FLAT.

It should be observed that the pumice all through this
region is very similar to that around Vesuvius. Leaving the
trail, and crossing a pretty little gurgling stream, we were
soon at the oldest destroyed camp, where there is scarcely a
doubt that Van Horn and his companions began to prepare
for winter quarters in 1862, when the Indians prevented
them. The thought and its associations were truly thrilling!
None can stand there, knowing the history of that spot, without experiencing a genuine sensation.

It is a place for mystery. The stillness of death reigns
there, and for miles around. Neither there nor for several
miles of our route to and from that point did we see a living
animal form but ourselves and our horse. Not even a wild
flower cheered our pathway, though just west of the great
divide we had crossed the most beautiful columbines, lilies,
geraniums, larkspurs, Solomon's seals, gentians, lupins, castil-
lejas, alliums and other flowers were blooming in abundance.

There were a few thinly scattered patches of bunch grass, which the pony greedily enjoyed. Except a few scrawny bushes at intervals, this grass was about the only undergrowth that at all broke the whitish, sterile, uniform color of that pumice-covered flat and the adjacent mountain sides. Not a bird, a rabbit, a squirrel, a chipmunk, not even a lizard was seen; not an insect, except a few black ants, about a quarter of an inch long, upon the trees. These we found even at hights of 12,000 feet in those mountains, the only signs of animal life we met at such altitudes. Not even a mosquito appeared, though they usually extend to our snow line, and, on the opposite side of the summit ridge, two miles distant, they are thick along the San Joaquin, especially around a fine soda spring. In the latter locality deer also abound. The whole region, however, is covered with a dense forest of fine firs, spruces, an occasional yellow pine, and the handsome, dark-foliaged yew tree of California (*Taxus brevifolia*); very much like its noted and scarce English or Irish congener. The wood of the latter is used by the Indians for bows. Some of these largest firs are from 80 to 100 feet high, and three feet through. The yews gave a funereal appearance to this region, and well they might, for, within a few rods of the Van Horn camp, the later made brush camp and the spots where the two skeletons had been found, were pointed out. But, no Indians are there now to annoy or endanger you. We did not even see or feel their spirits, guarding the weird spot. The Van Horn camp is marked by several felled trees and their stumps, very old and about the size needed for building cabins. Near by also stand three white spruces, or firs (*Abies taxifolia*), a foot or two in diameter. Together they form a triangle. They are but a few feet apart. Deep into their bark the Indians have cut

SOME OF THEIR RUDE FIGURES,

Evidently as a record of an important event that occurred there. These figures are sometimes spoken of as crosses. But close inspection shows that one is that peculiar form of a man with a hat on, and the other of a man lengthened into a lizard, as it were, that abound among the hieroglyphics of the "Painted Rocks" of Arizona, and which were made, no doubt, by some branch of the Utes to whom the Pah-Utes of Owen's river are known to be allied.

Not far from these camps is abundance of red cement along the base of Red Cement hill. Somewhere near is supposed to be the famous cement mine, now entirely covered with pumice, and concealed from view, as related in a former chapter. The hight of this locality above the sea level is not far from 9,500 feet.

After carefully examining this interesting and really historical ground, so curious as a study in wild nature, without reference to its odd traditions or to the supposed mineral wealth that lies buried in the immediate neighborhood, we ascended the slope of Red Cement hill, and, from a commanding point, overlooked the country northward, towards Mono lake and around Gid Whiteman's camp.

We passed several claims that have been duly located within the last year or two. On the northern slope of this red volcanic cone or crater we found deep beds of snow. Just beyond this, among the firs and well up the mountain side, Mr. Wilson selected a spot for the promised blast. It was in a bed of black, spongy lava, near which lay numerous bowlders, of every size, up to many tons weight, of the red lava that forms the top of the cone. Small fragments of black, glassy obsidian were also found occasionally. While the work of drilling, charging and tamping was going on, and we gazed at and chatted occasionally about the obsidian mountains to northward, and other peaks and ridges to northwest in Prescott district, there came to our ears slight sounds, in the midst of a stillness otherwise profound. They came apparently from a point west of us, beyond a narrow gorge, under a steep, reddish brown cliff, between a quarter and half a mile away. Tap, táp, tap.

LISTEN!

There it is again—tap! tap! tap! tap! It is the distant click of a hammer upon the head of a drill! Some one is at work there prospecting for gold, but he does not show himself or hail us, yet he must hear us. My friend, the veteran cement hunter, looks up, and with a mysterious nod or two, "There," said he, "is a cement hunter secretly looking for it now, you bet your life!" No doubt it was. And how with us? Were we hunting and yearning for it, too? We may as well "acknowledge the corn." My comrade had been seeking it for fourteen years, and could the writer have found such a gold bearing ledge, it would have made him, with a lot of the very best of people, happy for life. But never let the "*auri sacra fames*" get the best of us! This was the last "cement hunting" on record; yet, no doubt, many have quietly but diligently sought it since, as some will do again next summer and for years to come, until many more valuable veins are found. The blast was soon fired, quite a hole was made, and the queer, vitreous, porous, dark lava was carefully examined, but no sign of gold there. Perhaps it is just as well so. That statement, "It is death to the finder," does not whet the appetite much for it, anyhow, does it? even if we don't believe the threat of the spirits.

The bright July afternoon was fast wearing away. Then came a hearty hand shake and

GOOD-BY;

A solitary ride, eight or ten miles over a dreary waste, and by trail through the dense forests to Mammoth City; five miles more through Pine City and among the beautiful lake sources of Owen's river; over Pumice gap and the great Summit ridge of the Sierra Nevada to Pott's meadow—most of the last ride after dark, along the soft pumice trail now familiar to horse and rider. How sound and dreamless was that good night's rest for a weary body, rolled in blankets on the ground, side by side with a hardy mountaineer, under a thick spruce brush shelter!

Next morning early, French's saddle train was joined for Fresno Flats, about fifty miles a little south of west of us, and soon over the rugged, winding mountain trail we were leaving far behind us, but never to be forgotten, the grand, wild, mysterious scenes of the Owen's River War and the Cement Hunters.

Now, readers, you have the facts. While it is not here asserted that all the incidents related of this cement hunting are positively true, the fidelity with which the most authentic traditions connected therewith are recorded in this sketch, as told to the writer, is vouched for.

N.B.—If readers detect error of date or details in any part of this descriptive and narrative sketch, the writer will deem it a favor to be so informed by letter, through the office of the Post, for correction in any reprint.

J. W. A. W.

⟶⋇ Three Historic Maps ⋇⟵

Showing the Headwaters of the Owens River, Legendary Location of the Lost Cement Mine

Colton's Map of California, 1858
Whitney's Map of California and Nevada, 1873
U.S.G.S. Topographic Quadrangle Maps:
Mt. Lyell, 1901, and Mt. Morrison, 1914

Sections of these maps are reproduced on the pages that follow.

Location of the headwaters of the Owens River, eastern California

❖ THREE HISTORIC MAPS ❖

One striking feature of California's early maps, even the best of them, is the empty white space east of the Sierra Nevada—no settlements at all, hardly any landmarks. Had it not been for the discovery of gold and silver ore there, that white space doubtless would have remained blank far longer. The harsh, high desert of eastern California had little to offer pioneer farmers, compared to the milder, well-watered lands west of the Sierra.

The first map (Colton's, 1858) shows graphically how very little was known about the eastern Sierra during the first years of cement hunting. The two later maps then show how rapidly white men peopled the region, and how soon their surveyors located and named the landmarks and mapped the topography in ever greater detail to larger and larger scale. The maps may also help you imagine the formidable uncertainties the earlier prospectors encountered, with little to guide them but hearsay. Perhaps, too, they will help you follow the travels of James Wright in the summer of 1879.

COLTON'S MAP OF 1858
California, 1858. Published by J. H. Colton, New York. 11 × 13 inches, counties colored. Scale: one inch = 57.63 miles.

Colton's was an important and widely circulated map. It was first published in 1853, then reprinted with little change for a number of years thereafter.

The white space east of the Sierra crest reflects both the total absence of white settlers and the scant knowledge of the region's geography. In contrast to the western Sierra foothills that are dotted with mining camps, the eastern slope has not one camp. Only the Sierra Nevada, Mono Lake, Owens River and Lake are identified. The map shows the lands east of the Sierra as part of Mariposa and Tulare counties. Mono County was not established until

1861, Inyo in 1866. Colton's map will help you picture the route
and the difficulties encountered by the "original finders of the cement" (Original Story of the Cement Hunters), as well as the
route of "Kent and McDougal" from Stockton to the upper San
Joaquin River in 1869 (A Queer Episode).

WHITNEY'S MAP OF 1873

Map of California and Nevada, State Geological Survey of California, 1873. New York. 34½ × 41 inches, lightly colored along
state and county boundaries. Scale: one inch = 18 miles.

This important California map was based largely on the topographic material gathered by the State Geological Survey, under
the direction of Prof. J. D. Whitney, State Geologist. The State
Survey began its work in 1860. On this map areas of the state that
had never before been mapped are delineated in great detail.

East of the Sierra, Whitney's map shows not white space but,
instead, many named streams, mountain ranges and named peaks,
wagon roads and trails, settlements and mining camps. Benton,
Aurora, Bodie, Bridgeport, Monoville, even Dogtown—all mentioned by Wright—are new names since Colton's map. Prominent
Sierra peaks here are located and named for the first time. As-
suredly this is the map (or its second edition, 1874) Wright refers
to in the Introductory—"From Whitney's description and map I
take the highest of these to be. . . ."—as he describes the views
from the summit of Mineral Hill.

U.S.G.S. TOPOGRAPHIC QUADRANGLE MAPS:
MT. LYELL, 1901, AND MT. MORRISON, 1914

Topographic Quadrangle Maps: Mt. Lyell (1901, surveyed 1898–
1899) and Mt. Morrison (1914, surveyed 1911–1912), U.S. Geological Survey. Scale: One inch = 2 miles. Contour interval, 100
feet.

These two maps belong to the national topographic map series;
maps in this series are the base for all land maps made by others.
The U.S. Geological Survey began work on the series in 1879, in a
program to map the entire country, and has continued the never-
ending task of revising maps and drawing new ones to larger scale.

Compared to Colton's and Whitney's maps, these topographic maps reflect an enormous increase in knowledge of the land's geography. Benchmarks and contour lines give elevations for the entire region. Previously unnamed peaks and streams have been assigned names by the topographers.

Published many years after Wright's visit, these maps include ranches, a ranger station, and some place names that did not exist in 1879. Nevertheless, these maps are useful for locating many of the sites Wright described and for plotting the routes he likely took. On the maps the crest of the Sierra, "the great summit ridge" dividing the Owens from the San Joaquin drainage, coincides with the Mono-Madera county boundary. One error: *Pine City* mistakenly marks the site of Mammoth City. From all accounts, Pine City was about a mile farther to the southeast, near the shore of Lake Mary.

REFERENCES

Thompson, Morris M. 1979. *Maps for America*. U.S. Geological Survey.

Wheat, Carl I. 1947. "Twenty-five California Maps" in *Essays for Henry Wagner*. Grabhorn Press.

_____. 1957–1963. *Mapping the Transmississippi West*. Institute of Historical Cartography.

$\frac{1}{3,623,775}$, of Nature

CALIFORNIA

SCALE OF STATUTE MILES.

0   10   20   40   60   80   100

COLTON'S MAP OF CALIFORNIA, 1858
Courtesy, Central Map Collection, Stanford Libraries

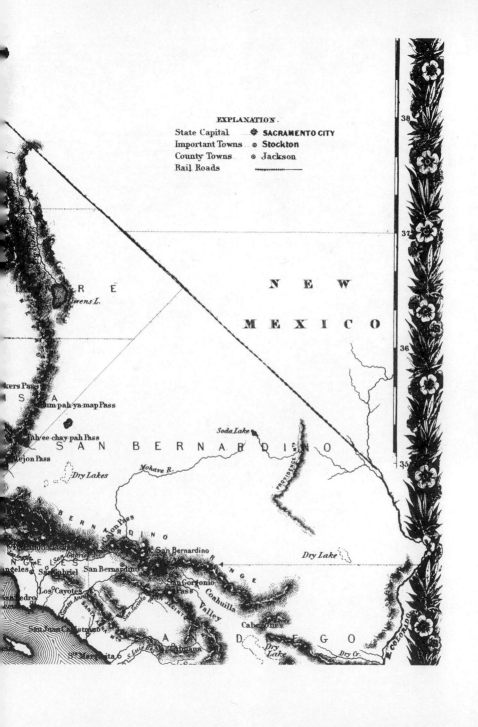

EXPLANATION.

State Capital ⚹ **SACRAMENTO CITY**
Important Towns ◉ **Stockton**
County Towns ◉ **Jackson**
Rail Roads ▬▬▬▬

N E W

M E X I C O

S A N   B E R N A R D I N O

Owens L.

kers Pass
um-pah-ya-map Pass
ah-ee-chay-pah Pass
lejon Pass

Dry Lakes

Soda Lake

Mohave R.

PROVIDENCE

B E R N A R D I N O

lon Pass

San Bernardino

Dry Lake

N G E L E S
ngeles   San Gabriel   San Bernardino
Los Cayotes
edro
anta Ana
San Juan Capistrano
San Gorgonio
Pass   Coahuila
Valley   Cabezones
S A N   D I E G O
S. Margarita   S. Luis   Tijuana
Dry
Lak   Dry Cr.

R. COLORADO

38

37

36

35

WHITNEY'S MAP OF CALIFORNIA & NEVADA, 1873 (enlarged)
Genny Smith Collection

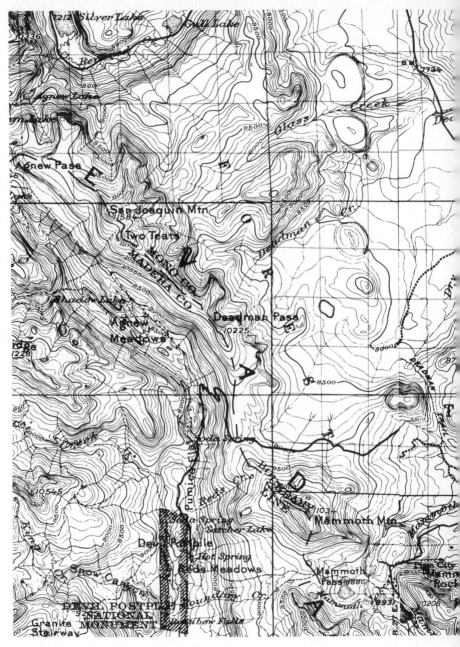

U.S.G.S. TOPOGRAPHIC QUADRANGLE MAPS *Left*: Mt. Lyell, 1901.
Right: Mt. Morrison, 1914.
Genny Smith Collection

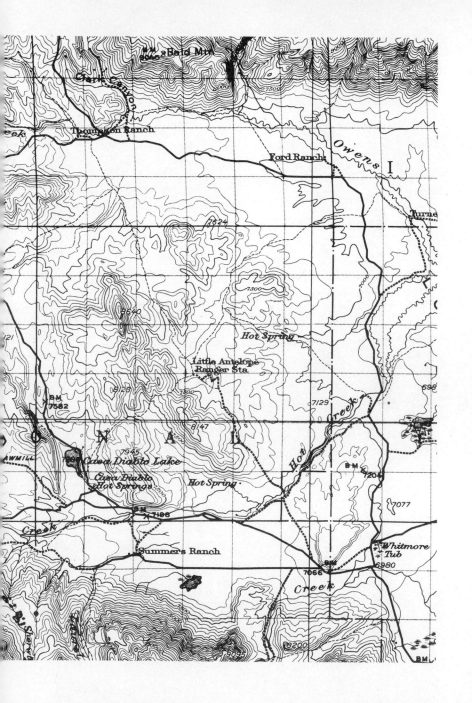

·✦ Appendix ✦·

A reproduction of Chapter XXXVII and illustrations from the first edition of Mark Twain's Roughing It, *American Publishing Co., Hartford, Conn., 1872. (Courtesy Stanford University Libraries, Department of Special Collections.)*

LAKE MONO.

·❧ The Whiteman ❧·
Cement Mine

IT was somewhere in the neighborhood of Mono Lake that the marvellous Whiteman cement mine was supposed to lie. Every now and then it would be reported that Mr. W. had passed stealthily through Esmeralda at dead of night, in disguise, and then we would have a wild excitement—because he must be steering for his secret mine, and now was the time to follow him. In less than three hours after daylight all the horses and mules and donkeys in the vicinity would be bought, hired or stolen, and half the community would be off for the mountains, following in the wake of Whiteman. But W. would drift about through the mountain gorges for days together, in a purposeless sort of way, until the provisions of the miners ran out, and they would have to go back home. I have known it reported at eleven at night, in a large mining camp, that Whiteman had just passed through, and in two hours the streets, so quiet before, would be swarming with men and animals. Every individual would be trying to be very secret, but yet venturing to whisper to just one neighbor that W. had passed through. And long before daylight—this in the dead of Winter—the stampede would be complete, the camp deserted, and the whole population gone chasing after W.

The tradition was that in the early immigration, more than twenty years ago, three young Germans, brothers, who had survived an Indian massacre on the Plains, wandered on foot through the deserts, avoiding all trails and roads, and simply holding a westerly direction and hoping to find California before they starved, or died of fatigue. And in a gorge in the mountains they sat down to rest one day, when one of them noticed a curious vein of cement running along the ground, shot full of lumps of dull yellow metal.   They saw that it was gold, and that here was a fortune to be acquired in a single day. The vein was about as wide as a curbstone, and fully two thirds of it was pure gold.   Every pound of the wonderful cement was

worth well-nigh $200.   Each of the brothers loaded him-self with about twenty-five pounds of it, and then they covered up all traces of the vein, made a rude drawing of the locality and the prin-cipal landmarks in the vicin-ity, and started westward again.   But troubles thick-ened about them.   In their wanderings one brother fell and broke his leg, and the others were obliged to go on and leave him to die in the wilderness.   Another, worn out and starving, gave up by and by, and laid down to die, but after two or three weeks of incredible hard-ships, the third reached the settlements of California ex-hausted, sick, and his mind deranged by his sufferings. He had thrown away all his cement but a few fragments, but these were sufficient to

THE SAVED BROTHER.

set everybody wild with excitement.   However, he had had enough of the cement country, and nothing could induce him to lead a party thither.   He was entirely content to work on a farm for wages.   But he gave Whiteman his map, and described the cement region as well as he could, and thus transferred the curse to that gentleman—for when I had my one accidental glimpse of **Mr. W.** in Esmeralda he had been hunting for the lost mine, in hunger and thirst, poverty and sickness, for twelve or thirteen years. Some people believed he had found it, but most people believed he had not. I saw a piece of cement as large as my fist which was said to have been given to Whiteman by the young German, and it was of a seductive nature. Lumps of virgin gold were as thick in it as raisins in a slice of fruit cake. The privilege of working such a mine one week would be sufficient for a man of reasonable desires.

A new partner of ours, a **Mr.** Higbie, knew Whiteman well by sight, and a friend of ours, a **Mr.** Van Dorn, was well acquainted with him, and not only that, but had Whiteman's promise that he should have a private hint in time to enable him to join the next cement expedition. Van Dorn had promised to extend the hint to us. One evening Higbie came in greatly excited, and said he felt certain he had recognized Whiteman, up town, disguised and in a pretended state of intoxication. In a little while Van Dorn arrived and confirmed the news; and so we gathered in our cabin and with heads close together arranged our plans in impressive whispers.

We were to leave town quietly, after midnight, in two or three small parties, so as not to attract attention, and meet at dawn on the " divide " overlooking Mono Lake, eight or nine miles distant. We were to make no noise after starting, and not speak above a whisper under any circumstances. It was believed that for once Whiteman's presence was unknown in the town and his expedition unsuspected. Our conclave broke up at nine o'clock, and we set about our preparations diligently and with profound secrecy. At eleven o'clock we saddled our horses, hitched them with their long *riatas* (or lassos), and then brought out a side of bacon, a sack of beans, a small sack of coffee, some sugar, a hundred pounds of flour in sacks, some tin cups and a coffee pot, frying pan and some few other necessary articles. All these things were " packed " on the back of a led horse—and whoever has not been taught, by a Spanish adept, to pack an animal, let him never hope to do the thing by natural smartness. That is impossible. Higbie had had some experience, but was not perfect. He put on the pack saddle (a thing like a saw-buck), piled the property on it and then wound a rope all over and about it and under it, "every which way," taking a hitch in it every now and then, and occasionally surging back on it till the horse's sides sunk in and he gasped for breath—but every time the lashings grew tight in one place they loosened in another. We never did get the load tight all over, but we got it so that it would do, after a fashion, and then we started, in single file, close order, and without a word. It was a dark night. We kept the middle of the road, and proceeded in a slow walk past the rows of cabins, and whenever a miner came to his door I trembled for fear the light would shine on us and excite curiosity. But nothing happened. We began the long winding ascent of the canyon, toward the "divide," and presently the cabins began to grow infrequent, and the intervals between them wider and wider, and then I began to breathe tolerably freely and feel less like a thief and a murderer. I was in the rear, leading the pack horse. As the ascent grew steeper he grew proportionately less satisfied with his cargo, and began to pull back on his *riata* occasionally and delay progress. My comrades were passing out of sight in the gloom. I was getting anxious. I coaxed and bullied the pack horse till I presently got him into a trot, and then the tin cups and pans strung about his person frightened him and he ran. His *riata* was wound around the pummel of my saddle, and so, as he went by he dragged me from my horse and the two animals traveled briskly on without me. But I was not alone—the loosened cargo tumbled overboard from the pack horse and fell close to me. It was abreast of almost the last cabin. A miner came out and said:

"Hello!"

I was thirty steps from him, and knew he could not see me, it was so very dark in the shadow of the mountain. So I lay still. Another head appeared in the light of the cabin

door, and presently the two men walked toward me.   They
stopped within ten steps of me, and one said:
   " 'St! Listen."

ON A SECRET EXPEDITION.

   I could not have been in a more distressed state if I had
been escaping justice with a price on my head. · Then the
miners appeared to sit down on a boulder, though I could not
see them distinctly enough to be very sure what they did.
One said:
   "I heard a noise, as plain as I ever heard anything.   It
seemed to be about there—"
   A stone whizzed by my head.   I flattened myself out in
the dust like a postage stamp, and thought to myself if he
mended his aim ever so little he would probably hear another
noise.   In my heart, now, I execrated secret expeditions.   I
promised myself that this should be my last, though the Sierras
were ribbed with cement veins.   Then one of the men said:
   "I'll tell you what! Welch knew what he was talking about

when he said he saw Whiteman to-day. I heard horses—that was the noise. I am going down to Welch's, right away."

They left and I was glad. I did not care whither they went, so they went. I was willing they should visit Welch, and the sooner the better.

As soon as they closed their cabin door my comrades emerged from the gloom; they had caught the horses and were waiting for a clear coast again. We remounted the cargo on the pack horse and got under way, and as day broke we reached the "divide" and joined Van Dorn. Then we journeyed down into the valley of the Lake, and feeling secure, we halted to cook breakfast, for we were tired and sleepy and hungry. Three hours later the rest of the population filed over the "divide" in a long procession, and drifted off out of sight around the borders of the Lake!

Whether or not my accident had produced this result we never knew, but at least one thing was certain—the secret was out and Whiteman would not enter upon a search for the cement mine this time. We were filled with chagrin.

We held a council and decided to make the best of our misfortune and enjoy a week's holiday on the borders of the curious Lake. Mono, it is sometimes called, and sometimes the "Dead Sea of California." It is one of the strangest freaks of Nature to be found in any land, but it is hardly ever mentioned in print and very seldom visited, because it lies away off the usual routes of travel and besides is so difficult to get at that only men content to endure the roughest life will consent to take upon themselves the discomforts of such a trip. On the morning of our second day, we traveled around to a remote and particularly wild spot on the borders of the Lake, where a stream of fresh, ice-cold water entered it from the mountain side, and then we went regularly into camp. We hired a large boat and two shot-guns from a lonely ranchman who lived some ten miles further on, and made ready for comfort and recreation. We soon got thoroughly acquainted with the Lake and all its peculiarities.

THE MINER'S DREAM.

·⟶⟨§⟩· Notes ⟨§⟩⟵·

Wright's history of "The Cement Hunters" originally appeared as part of a longer serial entitled "The Owen's River War and The Cement Hunters" published in the San Francisco *Daily Evening Post*. This article was composed of nine chapters, published as follows:

Nov. 8, 1879, p2c1-2, *Chap. I*   Introductory
p2c2-4, *Chap. II*   The Owen's River Indians

Nov. 15, 1879, p6c1-2, *Chap. III*   Original Story of the Cement Hunters

Nov. 22, 1879, p2c1-3, *Chap. IV*   Owen's River War—Origin and Incidents
p2c3-5, *Chap. V*   The War Continued

Nov. 29, 1879, p6c1-2, *Chap. VI*   Episodes and Later Incidents of Cement Hunting

Dec. 6, 1879, p6c1-2, *Chap. VII*   The Cement Hunting Continued
p6c2-3, *Chap. VIII*   Late Visit to the Destroyed Camps

Dec. 13, 1879, p6c1-3, *Chap. IX—And Last*   A Queer Episode—Who Was He?

The eighth chapter was apparently intended to be the last, but at its conclusion Wright added, "Should another chapter on this subject appear, it will be a brief romance founded on facts connected with this history." The following week the ninth chapter, "A Queer Episode—Who Was He?" made its appearance. This chapter has here been made to precede the eighth in order to retain continuity.

1. Mount Parker was named for James A. Parker of Bishop, who in company with B. N. Lowe, B. S. Martin and N. D. Smith, discovered the nearby Alpha mine on June 20, 1877. This mountain should not be confused with Parker Peak, seven miles northeast of Mount Lyell.

2. Although Wright's elevations are consistently too high, as much as a thousand feet, his distances are quite accurate. See Note 6 below.

3. Mount Lyell's waters do not flow into the Owens River.

4. Owens Lake and River were named in 1845 by John C. Fremont in honor of Richard Owens, a member of his third expedition.

5. The "tall, white, sugar loaf cone" was renamed Mount Montgomery in 1917. The name *White Mountain Peak* was then bestowed on a higher (14,246 feet), dark gray peak 15 miles farther south.

6. There can be little doubt that Wright is referring to Whitney's map of 1873, see section on Three Historic Maps. Except for local names near Mammoth City, all place names Wright mentions are on Whitney's map. Also, distances and directions to points Wright cannot possibly see from Mineral Hill (such as Death Valley and

Mount Whitney) agree with the map. Did Wright have the map with him, as he stood there on the summit of Mineral Hill? Or did he obtain it later, at home in Hanford, when he wrote up his story?

Although Whitney's map includes much more information than previous maps, it gives no elevations. Such determinations were many years away. Until publication of the USGS topographic maps of the area, more than twenty and thirty years later, elevations were mostly guesswork. Whatever Wright based his guesses on, they are consistently too high by as much as a thousand feet. For example, the high granite peak he is standing on, which he estimates at 13,000 feet, is 11,700. Pumice Gap (Mammoth Pass), which he estimates at 10,300 feet, is 9290.

7.   Pumice Gap on the south side of Mammoth Mountain is identical with Mammoth Pass.

8.   Another version of the original discovery of the "Lost Cement" is given by Mark Twain in Chapter XXXVII of his *Roughing It.* See Appendix.

9.   So called for the abundant steam vents, hot springs, craters, obsidian domes, and recent volcanic deposits of pumice, black lava and red cinders.

10.   The county seat of Fresno County from 1856 to 1874. The reservoir formed by Friant Dam now covers the site of Millerton.

11.   Langley's *San Francisco Business Directory* fails to list any Dr. Randall in the city between 1856 and 1861, but from 1862 to 1865 it does list a Peter Randall, physician. In 1866 the spelling of his last name is changed to Randle and as such he is listed until 1872.

12.   In most cases when Wright mentions the *San Joaquin* or *upper Joaquin,* he is referring to the Middle Fork of the San Joaquin River that lies between the Ritter Range and the Sierra Crest.

13.   *Cement* was a term commonly used by miners of that day for conglomerate, a rock composed of gravel and pebbles naturally cemented together by clay, calcium carbonate or some other material. Conglomerates derived from stream gravels do occasionally contain gold. However, throughout this story Wright uses the term *reddish cement* interchangeably with *reddish lava.*

14.   The Monoville, or old Mono, placers were discovered on July 4th, 1859, by Cord Norst, a miner from nearby Dogtown.

15.   Gid Whiteman, referred to as Gid F. Whitman by Joseph Wasson in letters to the Bodie *Weekly Standard* in 1878, was probably the Gideon F. Whiteman, miner, listed in Langley's *San Francisco Business Directory for 1862.* Whiteman, or Whitman, periodically prospected the eastern slope until August of 1880, when paralysis forced him to permanently retire to San Francisco. There he allegedly died in 1883.

16.   Whatever this Australian specimen was, it was destroyed along
      with all the Academy's collections in the San Francisco earthquake
      and fire of 1906. The Academy of Sciences today has specimens of
      gold-bearing conglomerate ("cement") from the Sierra's western
      foothills, but none are reddish. Jean DeMouthe 1984, personal
      communication.
          The specimen of "gold-bearing cement" from Australia could
      well have been gold-bearing conglomerate (but not lava), which
      was mined in the state of Victoria beginning in 1852. Gold there
      occurred in "deep leads"—well-cemented conglomerates, in
      stream valleys, that had been covered by lava flows. Demetrius
      Pohl, 1984, personal communication.

17.   This tributary of the Owens River still bears the name Deadman
      Creek and gives its name to Deadman Pass near its head, five miles
      north-northwest of Mammoth Mountain. On November 22, 1879,
      one week before the publication of Wright's version of Hume's
      murder, the Mammoth City *Herald* published the following
      account:

          "Near this place (Deadman Creek) is where a man who pre-
      tended to know and be a partner in these rich diggings (the "Lost
      Cement"), murdered a stranger for his money after having enticed
      him into the wilderness under pretense of selling him an interest in
      the mines. Hence the name of Murderer's Fork. The murderer
      only got about $500 for his hellish deed. The stranger was well
      mounted on a fine mule, which the fiend took, turning his own
      broken down horse loose. He then rode rapidly to the old Mono
      diggings, or Dog Town as it was called, and which was then a flour-
      ishing placer camp. He reported being attacked by Indians and his
      companion killed, himself being shot in the leg. This was merely a
      flesh-wound, and it was afterward ascertained that he did it with
      his own pistol as there were no Indians in that part of the country
      at the time. Immediately on hearing the circumstances a party of
      twelve or fourteen miners hastily armed and mounted and piloted
      by an Indian well acquainted with the country, started for the
      scene of the tragedy. After a long search the body was found, minus
      the head, which was found some distance from the body under a
      shelving rock. The remains were hurriedly buried, and the lonely
      grave is still to be seen—a monument to man's treachery toward a
      confiding companion. In the meantime the villain, who had been
      left by the miners under guard in a saloon at the gold diggings,
      succeeded in getting his keeper drunk, mounted the mule, made
      his escape, and has never been heard from since. This is as near the
      story as we can arrive at it after conversing with several who say
      they were in Mono at the time."

          W. A. Chalfant in his *Tales of the Pioneers* gives the following ad-
      ditional particulars, gleaned from some unnamed source:

          "Farnsworth met a man named Robert Hume in Carson, Ne-
      vada, and learned that he had some money. Farnsworth showed

him rich quartz from a Mariposa County mine. He said it had come from the head of Owens River; if Hume would go with him, and would put up the money for a small mill with which to work the claim there, he could have a half interest. Hume accepted, and took with him $700 to be used toward buying the mill in Mariposa. Not long afterward Hume was found murdered and suspicion centered on Farnsworth, who thereupon disappeared."

18. Wilson was probably J. F. Wilson of Visalia, who according to W. A. Chalfant in his *Story of Inyo* "squatted" on land near Independence about August of 1861. He apparently later returned to Visalia, since the Mammoth City *Herald* of February 25, 1880, in speaking of his discovery of the Minaret District mines refers to him as Mr. Wilson of Tulare.

19. Prescott District included the Mono Pass area above Bloody Canyon.

20. You can trace their route on Colton's map of 1858, see section on Three Historic Maps. It shows Knight's Ferry southeast of Stockton, Snellings, Millerton, and the San Joaquin River heading northeast toward the Sierra crest. Not shown on the map are Crane Valley (today the site of Bass Lake reservoir), 20 miles northeast of Millerton, and Beasore Meadows, about 11 miles northeast of Crane Valley.

21. Homer District was in Lundy Canyon, west of Mono Lake.

22. Perhaps Pott's meadow is the same as Reds Meadow, named after Red Sotcher, whom Wright mentions. It was said that Sotcher raised vegetables in the meadow and sold them to the miners at Mammoth City.

23. Today's town of Oakhurst is at Fresno Flats.

⚜ Selected Reading ⚜
Recommended Maps

Most of the books and maps recommended below are widely available in Inyo and Mono counties. Along Highway 395, from Lone Pine to Bridgeport, museums, ranger stations, visitor centers, bookstores, and some sporting goods stores stock many regional books and maps.

Brewer, Wm. H. Francis Farquhar, Ed. 1974. *Up and Down California in 1860–64: The Journal of Wm. H. Brewer.* Univ. of California Press. 584 pp. Cloth and paper.

Browne, J. Ross. 1978. *A Trip to Bodie Bluff and the Dead Sea of the West in 1863.* Outbooks. Paper.

Chalfant, W. A. *Gold, Guns and Ghost Towns.* Chalfant Press. Cloth and paper.

DeDecker, Mary. 1966. *Mines of the Eastern Sierra.* La Siesta Press. 72 pp. Paper.

Smith, Genny Ed. 1976. *Mammoth Lakes Sierra.* Genny Smith Books distributed by Wm. Kaufmann Inc. 148 pp. Paper.

RECOMMENDED MAPS

Inyo National Forest Map, U.S. Forest Service, 1979. 30 × 40 inches, color. Scale: ½ inch = one mile. Covers the entire Inyo Forest from its boundary south of Owens Lake north to Mono Lake. Road and trail information is more reliable than older maps. For sale at ranger stations and at the Supervisor's Office, Inyo National Forest, Bishop, CA 93514.

U.S. Geological Survey Topographic Maps, 15-minute series. 18 × 22 inches. Scale: one inch = one mile. The Devils Postpile and Mt. Morrison Quadrangles (both 1953) show the headwaters of the Owens River. Purchase by mail from: Western Distribution Branch, U.S.G.S., Box 25286, Federal Center, Denver, CO 80225.

Geologic Map of the Devils Postpile Quadrangle, Sierra Nevada, California by N. King Huber and C. Dean Rinehart, U.S. Geological Survey, 1965. Map ˙GQ-437. 28 × 41 inches, color. Scale: one inch = one mile. Purchase from U.S.G.S., address above.

·✦ Other Books ✦· on the Eastern Sierra Published by Genny Smith Books

Mammoth Lakes Sierra: A Handbook for Roadside and Trail
Fourth ed., 1976. 192 pages, paper.

Deepest Valley: Guide to Owens Valley
Revised edition planned for 1985.

Owens Valley Groundwater Conflict
P. H. Lane and A. Rossmann, 1978, 28 pages, paper.
Critical issues of the Inyo County lawsuit to limit groundwater pumping for the Los Angeles Aqueduct.

Earthquakes and Young Volcanoes along the Eastern Sierra Nevada at Mammoth Lakes 1980, Lone Pine 1872, Inyo and Mono Craters
C. Dean Rinehart and Ward C. Smith. 1982, 64 pages, paper.

Old Mammoth
Adele Reed. 1982, 194 pages, cloth and paper.
Stories and memories of Old Mammoth and the Mammoth gold mines. Superb collection of historic photographs.

Doctor Nellie: The Autobiography of Dr. Helen MacKnight Doyle
Foreword by Mary Austin. 1983, 364 pages, cloth and paper.
One of California's first women doctors writes vividly of life in Owens Valley and its mining camps from 1887 to 1920.

For prices and mail order information:
William Kaufmann, Inc.
95 First Street, Los Altos, California 94022

Design by David Mike Hamilton and J. Wilson McKenney.

Text Linotype set in Caledonia by Wilmac Press.
Front and backmatter set in Linotron Caledonia by G & S
Typesetters, Inc.

Headings phototypeset in City Medium and City Light by
G & S Typesetters, Inc.

Printed offset by Braun-Brumfield, Inc.

Bound by Braun-Brumfield, Inc.

VIVIANE AMAR

*Chère Sylviane,
In appreciation of your
friendship,*

Children of the
Universe

Fable for the Future

Viviane

Illustrated by LeeAnn Pollak

http://www.leading-leaders.com

Contents

Dedicated to my mother, my father, and my ancestors;

To my descendants, from generation to generation;

To us, our families, and our children;

To the universe that joins us together.

Foreword by Albert Jacquard

We rarely rest our gaze upon even the most elaborate of cathedrals when we pass by it routinely every day. We no longer pay attention to the most ingeniously organized arrangements, as long as they operate exactly as expected, day after day. The result is that we cut ourselves off from a primary source of personal enrichment: our capacity for wonder.

If we took a good look at ourselves, that capacity could thrive. Yet when we think of how our bodies work, we're rarely filled with wonder; we simply assume it's all a matter of routine.

To help us gain a sharper perspective, Viviane Amar has envisioned a drama. Ignored and disdained, our cells remind us of how vital they are by going on strike, or rather by staging a series of strikes that affect various organs in succession. As the cells in our lungs, digestive tract, and brain take turns in refusing to do their jobs, the whole delicate, fragile machinery that each of us is moves closer and closer to breakdown.

Luckily, it all works out in the end. But along the way, the author delivers a compelling lesson in biology to lay readers, and an equally powerful lesson in communication to biologists.

With this book, Viviane Amar has bestowed upon us a priceless gift. She has shown us how to restore our sense of wonder.

A.J.

Geneticist

Prolog

Humanity is about to reach a tipping point in its history. Due to the irreversible effects of our acts, our collective fate will take a disastrous turn unless we react immediately. Our choices bear witness to a paradigm of destruction, a propensity for death rather than life. We destroy ourselves through what we ingest or absorb (food, air, medication with unwanted side effects—to mention but a few). We dehumanize ourselves by relating more to machines than to other human beings. We decimate our planet, out of indifference and apathy. And we are engaged in a monumental battle between barbarism and civilization, pitting hate against love. No matter many dire warnings we hear, we still fail to react.

In predatory fashion, human beings disrupt nature and her treasures, unbalancing cycles, stealing precious time from our planet, casting our fellow humans and animals into turmoil.

Our African birthplace, once a Garden of Eden, where the children of the Maasai—proud cousins of our ancestors—still leave bare footprints in the dust has become a vast desert. Many of the babies born in Africa and Asia won't make it to age one. Their parents are waiting for a helping hand in vain.

Hundreds of billions of cubic meters of groundwater are extracted every year. Our drinking water contains hundreds of pollutants and infants' umbilical cords are contaminated by as many as two hundred chemicals, heavy metals, and pesticides. And what do those pesticides help grow? Only the bank accounts of those who have succeeded in monopolizing humanity's resources. Water and food are life, medicine; they have been turned into killers.

The tropics and their inhabitants suffer more and more each day from the climate change we have done so much to bring about. Landslides, floods, and the drying up of once vast lakes have become commonplace occurrences. The Greenland Ice Sheet is shrinking almost visibly; icebergs the size of Manhattan melt away from one day to the next.

A century of industrial development has turned our fertile, fruitful Earth into a wasteland. Which animals should we save from extinction? That is the kind of question now on our grim discussion agenda.

With the same blindness, the same hatred of others and of themselves as ever, brothers resume their mortal combat as soon as the slightest resentment surfaces. But how can you not be a stranger to the Other, the Earth, the Universe, when you are a stranger to yourself?

The crucial question, however, is how I can get in touch with myself and my fellow creatures.

I spent a long time searching for God on the Moon—until the astronauts landed there. Scientists may go as far as they like, trying with their space probes to learn even more about the origins of the Universe than electronic eyes can detect. The discovery of the famous "first second" may overjoy them and their neurons. They may be driven by the pursuit of an equation that will reduce all unknowns to a single, elegant formula, providing the explanation for gravity and the quantum world, for the primitive atom and dark matter, for dark energy and hungry black holes. Yet I suspect the only answer I could give the conjurer within me to the question of how life has evolved is that it is a Mystery.

In the infinitely large—our universe—and the infinitely small—my galaxy—in the even smaller "Who am I?" and the invisible surrounding me, I keep coming back to the big question: "Where do I belong? How should I understand my passage between these two infinities?"

In any case, I have a utopian vision. I believe that tales, stories, and fables can raise awareness. My hope is that a quest for the wondrous in ourselves, in our planet and our universe, will bring us closer to the meaning of what we are experiencing and will ultimately provide insight into how we should act.

So a story began to take shape in my mind. I have set out to discuss in the simplest possible manner what is universal in our universes. My aim in telling this story is to convey a sense of the wondrous. My hope is that everyone will love our Earth enough to want to protect it. My dream is that rather than the death of their fellow creatures, human beings will prefer Life.

1

The Revolt of the Cells

2025. Storms and tornadoes devastate the Earth, leaving eight billion individual universes sprawled limply on the ground, mouths hanging open, all color drained from their faces. Humanity has been bled dry. Having sacrificed its most precious resources and squandered its inheritance, humanity is now dying. Desolation is everywhere, from people's hearts to the gaping craters where forests and taiga once stood.

Realizing they can no longer trust *Homo sapiens*, the four cell colonies living in each human being resolve to take matters into their own hands. The year 2025 thus begins with a revolution the likes of which no one has ever seen or imagined. In four successive waves, the insurgents lay siege to the humans.

What the unsuspecting humans do not know is that every one of them is highly populated. Since the intricately structured cells existing within them were first discovered, it has become clear how extraordinarily efficient and endlessly dynamic those cells are. They continually evolve toward greater complexity, adapting to both major climate change cycles and their immediate environment—and they do so in a seemingly effortless manner. Like galaxies composed of millions of networks, cells don't just transmit information; they connect up to flash each other messages that celebrate emotion and reason. To convey what our species is all about, its fears and its aspirations. To enable humans to relate to each other.

But for several decades now, the world created by humankind has been wearing the cells down. Deaf and blind to the sublime within them, in exile from themselves, people have become estranged from each other and indifferent to the fate of their planet.

In response, the cells, differentiated into four main functions reflecting their distinct contributions to life processes, now organize into four powerful divisions and declare war on humanity. That's right, war—the language that all human beings understand! Rejecting alienation, the cells embark upon four strikes that will soon leave the sons and daughters of *Homo sapiens* badly shaken.

Each cell function forms its own coalition:

- The Alveoli, the Divine Breath that brings the body to life and delivers its underlying rhythm, initiate the first strike.

- The Mucous cells in the dark chamber of the human belly, cells of digestion which transform food into the energy that life requires, stage the second walkout.

- The Neurons—cells that think and act, love with all their senses, and lay the foundations for the worlds to come—are the third group to "down tools."

- The countless Spermatozoa—hyperactive cells programmed for reproduction—and their single Ovula companions are the last to rebel. This fourth strike will spell the end of human life.

As the cells see it, our species titled itself Humanity without becoming fully human. The era of *Homo sapiens* is drawing to an end, because the human race has shown too little wisdom. The very beings who dream of traveling to the stars and encountering extraterrestrials lead grimly solitary lives. Often engaging in barbaric violence, female prostitution, and child exploitation, they have wrecked their unique planet, along with the skies above it..

After sounding the alarm too many times to no effect, having experienced discouragement and sadness (the sister of anger), the cells feel the time has come to take a firm stand—to speak out and to act. They thus convene a High Council in this somber period of the year 2025, as the lament of the winds fills the firmament. In the strange buzzing hum characteristic of their chemical and electrical language, inaudible to all those who fail to plumb the depths of their hearts, the cells issue the following statement to the humans:

> *From father to son, from mother to daughter, we have always conveyed our knowledge and memory to you by means of the lymphatic fluid surrounding us. In this inner sea, which calls to mind the primal ocean of our origins, we are like amphibians without diving suits, permeable to our environment, created to create you. We are unlike anything else.*

As citizens acting in solidarity with the singular world that each human being forms, as comrades with a shared destiny, we know in the depth of our beings that at the core of both microcosms and macrocosms is relationship.

Since well before the time of our distant ancestor Toumaï and our African aunt Lucy, we have been working to impart greater awareness to you, so that your decisions, your words, and your hearts will be right. We constantly take part in shaping your identities, rewriting your biographies in accordance with your culture, your memories and your desires, your goals and your emotions, your fears and your dreams. For each human is unique. And each human's truth, all too often brandished as an absolute weapon, is but a facet of many ephemeral prisms.

Neither perfect nor invincible, peace-loving but not passive, our courageous friends the T cells are locked in incessant struggle against all aggressors, unbeknownst to you. For millions of years, they have been successfully forming powerful alliances with the majority of bacteria. A fever sometimes occurs to let you know that we are protecting you. What makes the immune system so strange is its awareness of both self and Other; a consciousness that feels and heals.

The cells are interrupted by a network of fabulous lymphocytes that are more fearless than others. Proudly and in unison, they add:

By strengthening your immune system, we make you smarter!

Not the least bit annoyed by this contribution, the cells resume:

Saving you means saving what makes Earth the most miraculous planet in the universe. But you cannot bring yourselves to recognize this, because with knowledge come love and responsibility.

For four billion years, we have never stopped loving you, even back when humanity was yet to come and everything was infinite chaos, spiraling into order. Of course, there was a second—at the beginning of the world—in which we were gas and magma, particles and molecules in a violent birth process. A tumultuous whole expanded until life appeared on the planet. Those who like to quantify the magical and the inexpressible will tell you that there was one chance in sixty-four million that the Earth—due to its shape, size, position in relation to the stars and its sun, its moon, its axis of rotation, the composition of its core and crust, its oceans and volcanoes, the energy created by its metals and its magnetic field—would become the Garden of Eden. And would give rise to another miracle, the miracle of a creature that both bears witness and acts, a creature whose complexity we have barely begun to fathom.

The Earth seems to have been created so that you could be the human beings that you are, with eyes and hearts, with all the sensations that move you, to encounter nature and everything it has to offer; its mountains and lakes, boulders and geysers, fruits and flowers, the thousands of animal species who are cousins to your fathers and mothers.

Thus, since time immemorial, we have stored the emotional heritage of survival in your brains, the memory of those overpowering primal fears experienced in a hostile world. But your emotional legacy also includes aspirations to something greater than yourselves. All of you once found yourselves at the foot of Mount Sinai, consciously listening to the ethical message. That was how you discovered empathy, and as oxytocin and dopamine, the hormones of love, got flowing, they led you to what best warms your hearts. Compassion for the Other is in your genes, your hormones, your entire bodies; it inhabits the Infinite.

Once again, the High Council gets disturbed, this time by a horde of fusiform cells that process information between the hemispheres of the brain, reconcile paradoxes, and nurture decisions made by the prefrontal cortex, a hotbed of intelligence and synthesis:

Don't forget to tell them that—no, wait, let us tell them. Humanity emerged in three stages. In the first one, its members were comparable to caterpillars, with their prospects limited to hunting for food and having one-night stands—at any price. Eros and aggression ruled supreme. In the second, cocoon-like stage, you advanced from aloneness to a sense of community, partnership, and good will, while continuing to struggle with male paradigms for managing territory and competition. In the final "fullness" stage, human beings become themselves through encounters with others, assuming their rightful place in the Universe. Returning from your exile from yourselves, you will fly off to those others like a monarch butterfly whose consciousness expands as it discovers new horizons

The cells nod in unanimous approval, before asking the countless networks to be so kind as to refrain from further interruption and allow them to finish.

As your cells, your proteins, your molecules, we must impress upon you that you are your environment. Without it, you can't eat, your hearts won't beat, and your genetic code can't even be reprogrammed. Your environment bombards you with messages of light and resonates within you. Do you hear that song from the cosmic microwave background of the Universe, in which your globe of humus and rock floats? It tells of the relational energy fields in which you are immersed and that you, in return, affect.

Today, however, we are extremely worried—about you and the Earth. The Blue Planet is warming up at great speed. Obsessed with their egos and short-term profits, faint-hearted governments and industrial barons respond with untrammeled power to the powerlessness they have created. Those who govern us, although elected by the many, have gone so far as to establish markets on which permits to pollute this unique paradise can be traded.

The Earth was a generous host to you during your adolescence. Do you really want to sacrifice it? You have the power to change everything, your relations, even your perceptions, in order to recreate yourselves. A new Covenant is possible. The time has come for all times to coalesce, human time and cosmic time.

Ever since your advent, we cells have sought to act humbly and fairly, whereas you have displayed complacency and arrogance. Your blindness apparently leads you to believe that the fleeting moment each of you spends on Earth has greater value than the course of all generations, than history itself. Aren't you forgetting the billions of years of change it took for humanity to come into being?

The sixth planetary extinction is mostly the work of Humankind, marking the first time that inhabitants of the Earth destroy it, devastating its jungles and paving over its valleys. Oceans once teemed with life, but now, entire schools of fish killed by mercury poisoning wash up onto beaches where children no longer play. Their lifeless bodies seem to cry out: "I used to be alive, vibrantly alive!" Misshapen frogs in mangroves foreshadow a future of uncontrollable mutation. The North and South Poles are in peril; the Ice Sheets collapse under the weight of bears and their young. Walruses and emperor penguins weep. The snows are no longer there forever.

In the arid lands of the South, where sand dunes are continually reshaped, famished elephants abandon the scorched savannas and lay villages to waste. Millions of human beings languish, although affluence is just a stone's throw away. Mothers sob tearlessly while others rend the universe with screams inaudible to those who have satisfied their wants. The dispossessed are denied food, heritage, and culture. And though they often rise up in anger, the privileged remain indifferent.

Humans, did you say? Humanity has no meaning unless it is shared. With their backs turned to themselves, oblivious to the call that caused them to stand tall, people know themselves least of all. Sleepless nights and melancholy have become the fate of the majority. Is this an indelible trace of the nirvana that preceded the first cell division, a remembrance of when sperm and egg met in a womb of blinding light, a longing for an exhilarating, life-giving encounter? Or is it sorrow over not being immortal? Either way, our planet is a lonely place.

Humans communicate with another like unhappy children, particularly the angriest among them. Seeking recognition, individuals strive to be seen and heard. Narcissus has gone astray; did he ever really love himself? Facing inward, he suffers feelings of inadequacy. He no longer knows how tall and handsome he is. Aiming to please, fearing vulnerability and authenticity, he dons a mask, yet finds himself constantly judged. This unloved being roams his ramshackle inner palace, no longer glittering like the Koh-i-Noor diamond he is. There is no one to tell him he will eventually become himself. Escapism, gloom, sickness, vengeance, and war define his existence.

There are eight billion of you solitary beings, yet deep down inside of you, the urge to multiply spills over into a desire for encounters; a quest for the wondrous.

No savior will be coming, neither spirit nor living creature. Gone are the prophets who once exhorted you to become your own inner angels, who beseeched fathers and mothers to become the bearers of a new civilization. They were powerless to save you from yourselves. Will you be creatures of division, or of multiplication? Imagine, O! heart of mine, eight billion human beings at peace at last!

The cells have shared many of their secrets with wise and inquisitive people, hoping to be taken seriously: to no avail. Stunned at first, the cell communities are soon beside themselves with anger. Powerful waves run through their hosts, causing them to falter. Oceans rise up above blue sharks. Thick plankton engulfs the seabeds.

The cells still hope that humans will see the light before the start of the strikes they are planning. But nothing happens. In unison, they chant:

We continually create you in accordance with an irrevocable program that each of us stores in our memories, even as we maintain our ability to evolve. With mind-boggling precision, we draft maps of the locations and feelings within you, generating new neurons and nice, big, vibrant synapses that build you into different, unique, and constantly renewed beings.

You house the most incredible microprocessors—they are perpetually updated. You contain myriad clocks programmed for the development of humanized human beings. You represent the greatest spatial adventure, as both vessel and odyssey, trajectory and destiny. Yet you are reluctant to believe it.

Even the frailest cell has its own chemistry, its own guiding star. It also has its own brain—the membrane through which it perceives the environment, interacts with it, and on the basis of its interpretation formulates beliefs and expectations, becoming an optimist or a pessimist, shaping its view of the world and its own role within it.

Like the cosmos, you are inhabited by billions of suns. You radiate energy, yet you are apparently unaware of it. You wonder what, where, and how, while dodging the question of why, for you are afraid you will discover that human beings are more than just culture, psychology, neurons, and hormones. We have gone from being subjects to being objects of study that are scrutinized and dissected, although we are the builders and guardians of the short-lived temples that you are.

Some build with sand, stone, concrete, all too often with wind and ignorance. Like fireflies blinded on the planet Babel with its seven thousand languages, they abandon all moderation and strive for omnipotence. As the Achilles' heel of this planet, humans wielding power threaten each other with economic, chemical, even nuclear war. Alliances are formed— between dictators. Meanwhile, the masters of the world religions, who are supposed to be preaching the love of God, fan the flames of hatred, seeking supremacy or revenge for past humiliation. The only word left on Earth is violence.

Sound the alarm! Some of us have already broken off all dialog, and ignoring their neighbors, have gone from interdependence to indifference to annihilation, maddened by the folly of humanity. Beauty is becoming a beast. To make matters worse, we have no high priests among us and we don't always succeed in calming our fellow cells down. Their frenzy is contagious; how can we restrain it?

To be sure, more and more of you are acquainted with the music of the heart. Yes, men and women around the world are already fully human. The notes they produce are in harmony, inspired by the meaning of their journey and the sense their action gives. As beings of contribution, they reach out to strangers, who mirror their innermost selves. They stand tall, and when they look into the distraught eyes of others, they restore dignity.

In dusty villages, their hands build wells, roads, schools, and shelters. Through their labor, trees grow where pools had filled up with dry sand. They beat back disease and unsightliness, and travel the world in search of funds so that bountiful smiles may arise from the depths of wounded souls.

Upright beings find that uniting and cooperating leads to greater serenity than fighting and hoarding. Standing tall, and whether they lead infinitely small or infinitely large lives, they take a chance on sharing, so that other hands and eyes no longer need to beg. These beings versed in the excellence of the heart constitute the meridian of human consciousness. With and through them, Homo Filia *is born.*

But this fresh attempt by the High Council has next to no impact. As before, most friends are still virtual, while neighbors remain indifferent: connection has become a largely electronic affair, sadly enough even among the children. Thus, in the spring of 2025, all cells unanimously sing out:

We have told them about love and otherness, identity and interdependence, wisdom and maturity, life, but they show no sign of comprehending. What should we do?

— Let's try Reason!

— Have they ever possessed it?

— Then how about emotion?

— They are terrified of it.

The question is put to a vote. During this interminable instant of deliberation, cosmic tornadoes illuminate the planet. Palm, bamboo, and oak trees are uprooted. Photon cyclones merge with the storm created by the cell assembly. Black rain runs thick upon magnolias in bloom.

Across the Earth, anxiety takes hold. Could this be the end of the world? An eerie silence sets in.

2
The Alveoli Coalition

At this point, a deep voice pulses forth:

The humans show less wisdom than their children. We should go on strike, because most of them are unaware that everything they possess has been lent to them; their lives and their status, their work and their loved ones.

There are billions of solar systems, yet humans alone are living aerobic organisms. The first strike decreed by our High Council shall therefore involve respiratory cells, the Alveoli. And the first plague to hit the humans shall be a lack of air.

Without oxygen, all of us—plants and animals—would die together, except for a handful of bacteria at the bottom of the oceans. Over three billion years would then go by before human life would even stand a remote chance of reappearing on Earth. Our High Council hereby proclaims its first demonstration of force with exact timing. It shall last two minutes. Any longer and death will ensue.

The lesson is cruel indeed. The humans soon discover the implacable execution of this sentence. After so many years of illusory power, they are in for two excruciating minutes.

The countdown has begun. From the first few seconds, men and women determined to escape their fate start voraciously inhaling. The air they take in passes down to the bronchi, then to the bronchioles—only to be halted by the tissue lining the lungs. Panic ensues. The millions of alveoli who form this first coalition refuse to let oxygen enter the bloodstream, compressing the capillaries, those precious alveolar vessels devoted to transporting air to the body's organs.

In their clouded brains, the humans discern Alvea, the representative chosen by her fellow respiratory alveoli, as she mercilessly lays down the law:

Your body will just have to make do with lower supplies.

Masses of deadly carbon waste gas now prevent the diffusion of oxygen through the bloodstream. Poisoning progresses insidiously with every second that goes by. Mouths open, heads reeling, their hands held out in vain, the humans gasp for breath.

Alvea, Alvea, they implore her, Our cells are suffocating. Our supplies are running out. Please call off the strike, and fast! We get the point.

Standing by the group decision, Alvea counts the seconds. Medulla, the control room in the brain, calmly responds to the internal disruption. "This is no big deal; we will survive," she says to herself, and sends out the following instructions to the lungs, diaphragm, and heart:

Bring in some fresh air. Heart, there's no need to panic, so stop beating wildly—you'll only waste the oxygen you're getting.

The chest muscles, which routinely breathe in and out more than 25,000 times a day, appear crushed by an intolerable weight, and heave desperate convulsions. Medulla cries to the digestive tract for help:

I need sugar!

— My cells aren't getting the air they need to produce it.

The brain is on its last legs. Its neurons can live for more than a hundred years, but once they are damaged, they die. A frightened Medulla screams:

Get to work, blood pump! I'm suffocating, I'm dying—hurry!

Denied its life source, however, the heart is in the process of asphyxiation.

I'm begging you, heart. Don't slow down, implores Medulla.

A writer traces squiggles across a blank sheet of paper; an artist draws a child with a triangular eye; a sculptor shapes a mother with a hollowed, sunken belly. The millions calculated by an accountant become an empty string of zeroes, while a lover's hand dangles limply on the breast of his betrothed.

Under African skies, priestesses in grass skirts, their black bodies smeared with white kaolin, fall abruptly out of trances and slump onto parched ochre ground. A hunter's arrow wedges in a tree, to the relief of a gazelle.

Among India's green tea hilltop estates, kingfishers hover above women in multicolored saris who stand motionless, their baskets overturned.

Gazing toward the skies, the Lhasa hermit concludes that his time has come. "So this is how I am to die," he meditates. On Peru's mountain slopes, peasant women enjoying their usual Sunday soccer game collapse in a heap on their red-embroidered black dresses, as sheep and llamas look on in bewilderment.

Explorers on board a spaceship observe the Earth in powerless dismay. As they try to maintain contact with the mission control center from their cramped cabin, they are stupefied by the sight of engineers slumped over in front of their displays. Images of previous space travel disasters flash through the astronauts' minds.

An infinite world suddenly highlights how finite humanity is. Trains tearing across the Earth pull with a deafening din into strangely silent cities, carrying nothing but unconscious passengers. After struggling desperately to take in air through their skin like worms, thousands of souls have become motionless, for the oxygen they need to move, think, and feel is no longer available to them. As the planet sinks into lethargy, prayers and appeals fade before blue lips can even utter them. Sixteen billion eyelids close. The living are powerless to protect their children.

Auras fade away, as does planetary frequency. There is no one left to bear witness to human life on Earth. The last seconds tick by, with Alvea stubbornly counting down: 5, 4, 3, 2, 1, 0—nothingness.

At the last millisecond, Alvea orders her companions in her secret language to halt the impending demise.

Lung cells instantly reopen the channels of life, sending the good news to Medulla. Oxygen brings a revival of vigor. In successive waves, the humans open their eyes, part their lips, savor the light, touch their bodies as the flows running through them finally restore their former warmth and vibrancy. They shake their stiff limbs, smooth their hair. Laughter and the joy of breathing make their return. Pulses recover their normal pace, eyes their usual gleam, hands their precision, brains their clarity. Aromas and sounds awaken the senses. People purr sensuously beside their surprised cats.

A faint breeze casts away the shadows that obscured the stars a moment earlier. Could air be becoming a precious substance?

3

A Gut Reaction

Our species indulges in a long vacation in the year 2025. Before resuming their daily occupations, people set out to improve their environment. Trees are planted in deserts, parks are cleaned up, and forests are contemplated. There is widespread agreement on the use of new, less damaging sources of fuel. Young and old people alike congregate in gardens. The law of life returns to ecosystems.

Thousands of fragrances fill the air once again: basil, lavender, marjoram, curry, saffron. In the clear, warm waters of the Caribbean, the sunlight illuminates the dance of purple gorgonian seafans. In Tasmania, baby kangaroos eagerly suck on their mothers' teats. In the Tierra del Fuego, created when the oceans collided with the Andes mountain range, penguins frolic in limpid fjords, awaiting snows that have recovered their eternal character. Further north, on barren mountain plateaus, pumas charge flocks of vicuñas, attracting hungry condors.

To honor this renewed awareness, the humans arrange celebrations. In preparation for a gigantic feast, they pool millet and wheat, yew's milk from Greece, antelope from Kenya, bamboo shoots from Japan, chili pepper from

Barbados, vanilla from the Indian Ocean, kiwis and persimmons from Australia, Brazil nuts from Brazil, argan oil from Morocco, flowers from Bali, and shellfish from Tahiti. Now that Eden has been restored, all have become hosts and guests.

> *This is a day of celebration, says a woman elder. Let us dance. Censorship, lift your veils; fear, move on. Today I am receiving humans. My heart has reached its peak.*

As wine is served to seal friendship, sweet, throaty, or reedy voices break into melodies of old. Solos, choruses, the whole Earth becomes vibrant with song. The longing felt by humanity reaches the firmament as its music penetrates the infinite cosmos. The planet seems illuminated by the assembled humans, glowing with their ardor. Slowly, cautiously, animals approach these millions of sources of warmth and light.

As the night goes on, people begin to mistake excess for pleasure. Voices and bodies lapse into orgy; intoxicated by a haze of drugs, alcohol, and incense, adults neglect their children. The adolescents copy their parents, while the wise men and women who voice their misgivings are met with laughter and mockery.

The party is short-lived. Before the break of dawn, as rats and jackals, monkeys and cormorants rummage through the debris, the revelers' limp bodies are suddenly shaken by the onset of the second strike.

Rumor has it that the current strikers are the Mucous cells. Their very existence is unknown to most of humanity, although these are the most gentle, the most considerate, and above all the most protective cells in the body.

The stunned humans consult a student, asking in deliberately non-scientific terms:

Who, when, how, why? And for how long?

Showered with questions and delighted by the opportunity to share his erudition, the young researcher goes about telling the tragic tale of the Mucous cells, recounting the stages in their journey:

Did you know that before a first bite of food reaches our lips, millions of cells in our bodies get excited by the sight and smell of it? Or that these food lovers have a particular fondness for sugar and fat, which they use as fuel? Or that—

Warming to his audience, the student goes on to say:

Nature showed great foresight when it designed an ingenious technology to help the Mucous cells burn that vital energy. And who do you suppose plays a part in this whole setup for crushing, grinding, and liquefying food? The brain does, by manufacturing destructive acids with the deceptively innocent name of gastric juices that are so powerful they could digest one of your fingers if you sent it down into your stomach.

No one is amused by his cleverness, least of all the scientists who got sidelined so that a novice could hold forth. Which is what he does:

We don't feel hunger in our stomachs, but in our brains. We also contain gluttonous enzymes and precious bacteria that have been part of us for millions of years, which work like nuclear power plants. If they weren't there, we'd need

fireproof bodies to be able to withstand the incredible energy
unleashed by all the chemical reactions taking place inside of
us.

The student gets no applause. This second strike is starting to take effect in earnest, as a sour, acidic taste fills people's throats and eight billion bellies are racked with violent spasms.

Squat on their soft, yet solid bases, the brave Mucous cells decide it's time to speak up on their own behalf, because they know that speech is power. This is how they explain their hostile environment:

We're neither as elegant as the Neurons nor as flaming red as
the Alveoli. We have stocky, cylindrical bodies packed with
alkaline mucus that is the secret armor protecting us against
all those corrosive digestive juices. Without us, those acids
would burn you to death by perforating your stomach walls
with excruciating ulcers. Autodigestion is the horrible name
for that equally horrible outcome it is our job to prevent. As a
result, we have very short lives: a maximum of three days of
valiant struggle, before we get replaced as part of an endless
cycle. Our world is the stomach—a lair where a ferocious
beast lies in wait, like a cannibal demanding tribute. We live
in a world of decomposition to help you stay alive—

Before they can finish what they have to say, a major event modifies the course of this second strike: the other cells in the body give the floor to the digestive juices who have been charged with corrosion. Speaking in their own defense, the digestive juices counter the sympathy that the Mucous cells attempted to elicit:

We don't want to downplay the courage of the Mucous cells when they stand up under the steady stream of acidity we produce. But there is another side to the story as well. We break down food so that you can develop, and we draw life out of complex molecules.

The humans, however, are no longer listening; they are writhing in pain from stomach cramps. Each one suffers in his or her own way, depending on culture and experience, but they are all undergoing something unique. Outraged, frightened, or docile, they moan or face up stoically to the torture of inflamed mucous membranes. With dilated stomach walls and a bitter taste of bile in their throats, they huddle up as their bellies become the center of their being.

Come on; give me a break—I can't take much more of this.

The intensity of the pain leaves them at a loss for words. The bravest among them resort to folk remedies, get massages, consider hypnosis and relaxation. Others ingest huge quantities of pills that only make matters worse. Many cry, rage, throw tantrums.

Enough is enough. This is a nightmare!

— Then spare us the stress we're under, the exhausting excesses we are subject to. We are strong, but vulnerable. So show us a little respect.

At this point, the Mucous cells, who have taken quite a beating themselves, unanimously vote to end the second strike, and promptly begin secreting the mucus that shields them—and protects life. They industriously pour it into the millions of crypts in the intestines, steadily lining the walls with the sticky liquid. Determined to live, refusing to be cowed, standing firmly on their short legs, impressive with their stocky builds, the cells issue a thundering "NO!" that gives all bellies a powerful jolt.

The break proves beneficial to the humans, who rejoice as they rediscover what it is to have a body that cooperates.

4

A War of Nerves

Three months go by. The High Council has resolved to grant the humans a respite before initiating the third walkout.

One morning, the sun rises in iridescent blue against an orange sky. As the Earth takes on a purplish hue, dead trees begin to tremble. Baobabs suddenly lose their footing, and panic-stricken birds seek refuge in the reeds. Razorfish, skates, and sea turtles scurry to reach the depths. Darkened waters froth up around finback whales as they gobble down swirling schools of krill. Black-finned killer whales devour sea-lions on roaring beaches. As large snow geese look on, human time comes to a standstill.

A red cloud throbbing like a pulsating star beams out a resounding message to the entire world:

You have neglected me. You forget that I direct your lives. Using my precious Neurons, I constantly recreate you in your thoughts, giving you the ability to feel alive and believe you contain the Divine Spark.

That would be sufficient if you were just a collection of organs, functions, and systems—the respiratory, digestive, circulatory, and reproductive systems. That would be sufficient if you were only a network of cell networks, chains of neurons, dietary minerals, and smart hormones. That would be sufficient if you were nothing but archaic and cellular memories, episodic and long-term, selective and sensual. That would be sufficient if you were merely energy and electric, chemical, and nuclear circuits. That would be sufficient if you were only reason and emotion, both conscious and imaginary. And it would be sufficient if you were simply the outcome of Darwinian evolution, a singular ecosystem and the king of the animals.

But would it be sufficient to distinguish you from other species?

Is life just an array of probiotics capable of sustaining themselves, duplicating, and procreating? What meaning is there to this random chemistry? Why are there billions of individuals, billions of galaxies, billions of neurons? What present, what future is being served? Ever since you humans tried to kill God, you have been seeking to replace Him. You compute Him; you hunt for Him in the regions of the brain. You look for the sacred in your genes, for a phantom wandering the nervous system, and you reduce transcendence to neural networks and their electric and chemical interactions. How blind can you possibly be?

For millions of years, I have been developing myself to give you the ability to think for yourselves, to think in terms of mystery and Otherness, to conceive of and see beauty and purpose, the Good and the Just. Awaken your consciousness! The brain is what makes each person unique.

— But which one, the left brain or the right brain?

A good question indeed. Standing before an audience of neurons whispering to each other through their axons and ribosomes, in the ingenious tangle of countless dendrites, the left brain hastens to respond:

> *Communicating, speaking, reading, writing—that's what I handle. No one can outdo me when it comes to the logic worshipped by humans. I'm number one at certainty. I analyze and solve problems. I thrive on all things scholarly.*

> *— That's not true. I guide you in all your choices and decisions, in your fears, tears, and laughter, when you dream and when you give up. I am your inner language. Classification is essential, but so is an overall vision. I visualize your imaginary worlds, and I play a key role in intuition and creativity. I am where meaning vibrates.*

At this point, the Neurons scoff at both hemispheres of the brain. Like humans, the two of them have such a need for recognition that they emphasize their differences and ignore their complementarity.

> *You seem to be forgetting what makes you powerful—your interdependence. Isn't every memory and every decision laden with emotion and meaning? Isn't intuition the microprocessor of individual and collective history? And aren't they the products of so many thousands of years of cooperation? What would become of humanity if each of you stuck to your distinctive functions?*

Somewhat offended, the brain stammers in response. All the Neurons laugh, their serotonin-filled ribosomes ringing out like electric guitars. But this moment of laughter doesn't distract them from their commitment to launch the third labor action. Scheduled to go on for ten seconds, the strike is to be announced by Neurona, the brain cell union spokesperson.

Smoothing her impressive mane of hair, standing tall upon her axon channel—the nerve center sending out masses of information—Neurona makes full use of her superior communication skills:

> *We are going on strike to implement the agreement reached by the High Council, and to show you that we Neurons are your body's control center. We occupy a key position in the hierarchy, and our administrators report to the three highest-ranking generals in your individual kingdom: Cortico, Hippo, and Medulla.*
>
> *Close cooperation is what gives them their power. You contemplate the stars with your brain, not with your eyes. And there are things that you can see only with your mouth, your skin, your ears, your nose, your heart, and your soul. Yet there remains mystery. We are the first ones to quiver with pleasure when you receive a caress. Pleasure, by the way, is something we are particularly receptive to.*

Neurona takes a moment to flirt with a few loyal synapses, flooding them with gratifying dopamine, before resuming:

> *We set in motion millions of electric circuits that your scientists are good at recording. They've given them Greek names like Alpha, Beta, Theta, Delta, and after all, why shouldn't they? We are like an active beehive, with you as our queen. Our work continues until the end of your existence. So instead of saying, "It's a dog's life," you should really be*

saying, "It's a neuron's life." Nothing is accomplished without us: we're responsible for a woman's curves and the milk in a mother's breasts, for a man's beard and the size of his body.

You have highly specialized receptors for all your sensations. So our lives are devoted to capturing the outside world in order to feed your inner world. We transmit chemical and electric signals faster than a hundred yards per second, and we interpret them using powerful algorithms. Why? So we can nourish the emotional circuits that will develop more or less, depending on each person's own history. That's how the empathy or the anger networks are created.

We help construct the meanings that tell your personal story: what you say and what you forget, the space you take up and the space that you leave, the laugh you burst into or the affection that lacks, your blank gaze or the gleam in your iris, your sighs and your folded arms. Were you unaware that morality has ramifications in the heart?

She pauses briefly, and then continues:

Not to mention communication between one heart and another. Did you know that this is five hundred times more powerful than the relationship between the right brain and the left brain, which is the seat of reason? Did you know that the heart has its own brain? That with just forty thousand sensory neurites it emits magnetic waves that can be measured from a distance of miles? That it speaks to your brain more often than your brain sends it messages? That puts the heart ahead of both brain hemispheres, because it reacts faster than they do.

So the heart is an organ of awareness, one that is aware of its awareness and contributes to human awareness. It is aware of the flood of information reaching it, which it processes unfailingly. It sharpens your intuition, affects your emotions by the hormones it secretes, modulates its own rhythm and the pace of your breathing. Depending on what is going on, your heart energizes you, alarms you, or calms you; it can even foresee the future. The brain obeys the heart's wisdom, especially when souls bond together in the sweet waters of the hormone oxytocin secreted by the brain, which makes the one you love even more attractive to you. Gravitational pull connects you at astounding distances.

That's right: the field of love begins to operate in your hearts before the field of logic does, and it often drives you to bridge the unbridgeable in a relationship. To resolve the dilemma of both the conscious and the unconscious in our past or our instincts. A freer heart connects to the soul, communicating with the Universe, traveling across the ether, linking up spirit and matter.

Bashful lover of all things vibrant, the throbbing heart bursts out of its protective cage, beating wildly in its red Valentine attire. Impassioned by such flattery, it wants to declare its love. Neurona gestures gently to the heart to hold its peace. Swaying voluptuously on her long, diaphanous legs, eager to rejoin her companions, she briefly rearranges her hair, takes a firm stance, and then proceeds to say:

And be sure to tell your doctors who so barbarically dissect us that memory isn't lodged in any one location. It involves the heart, body, and soul, through billions of buried memories. It isn't a folder or an account showing deposits and withdrawals; it is a network of collective and individual pathways. The upshot is that your mental landscapes are all different, and without your knowing it, they all get renewed.

Scientists tell you that you have three kinds of memory, differentiated on the basis of how long they last. There is sensory memory lasting less than a second, short-term memory lasting a minute, and finally longer-term memory. They also claim that at the personal level, memories are formed through a process in which an incredibly swift radar system scans all the senses, and then marks each event and gives it an emotional coloration.

The body recalls pent-up sadness and anger, Neurona whispers. It remembers rejection, the striving for recognition, the visceral fear trapped in ironclad diaphragms, in jaws clenched to prevent words from being uttered. All these phantoms are struggling to break out of their crypts of affliction and embrace life.

You also know my views on this issue. By contributing to the development of meaning, memory becomes a heritage, a witness to the story of humanity. Its purpose is to create the future—a future of multiple possibilities. Not everyone can learn at the same pace; speed has no value as such. Setting direction is what matters.

My true belief is that there is also a universal memory we acquired as an embryo and that is everywhere present in the fields of energy of the universe. I believe this memory unifies time and space, beyond the nations' history and myths, beyond the individual highly emotional ones. The universe is memory of universal knowledge.

Yet how complex human resilience is! Thanks to us, you adapt and recreate yourselves despite your past. Humanity is the only species capable of self-becoming.

The plan to go on strike is nonetheless maintained, particularly because the powerful are still buying pollution rights from countries that take a more responsible approach.

You are about to have the ultimate experience, Neurona announces.

The work stoppage begins at dawn under a turquoise sky. Birds are still chirping, but all animal and vegetable species sense the strangeness of what is coming. Fauna and flora perceive human panic.

The flamboyant color of Royal Poinciana trees in Zaire fades, and the fragrance of Bengal roses undergoes a subtle alteration. Insects characteristically enamored of flowers suddenly fly away. The pheromones released by lionesses lose their power to stimulate rutting behavior, and the males anxiously withdraw. Eagles and peacocks— usually the proudest of all creatures—fold back their wings. The falcons of Arabia retract their claws. Out at sea, white corals cease to gobble up plankton. In fields, ants start to assist grasshoppers. In the forests of Denmark and Canada, dashing reindeer stop in their tracks. In the depths of the Antarctic, starfish group together, while nearby emperor penguins shoot out like rockets to reach the sapphire ocean. At the South Pole, snowy owls endowed with extraordinary night vision shut their big yellow eyes to keep from being blinded by a shower of light ions.

The whole planet is listening closely. Quickened by galloping fear, hearts are pounding out a new rhythm in unison. The humans know they are capable of the worst lunacy. Few of them are ignorant of the power and fragility of the brain, its genius and its madness.

The inevitable neuron walkout begins. Some people take a seat, while others lie down for fear of fainting. This is the frightening prelude to the ultimate experience. Many think of their fathers and mothers, of God, a lover, a favorite animal, an angel. Some let out muffled or unrestrained sobs. Sensing that this may be their last glimpse of the world, they hope to fill their souls with the rising or setting sun. Trembling in terror, they huddle together.

Cortico, Hippo, and Medulla order Neurona and her associates to stop sending all signals of the world and life. People run their clammy hands down their bodies, with the soothing yet alarming sensation that they are losing consciousness. All the impressions that ordinarily bombard the brain like cosmic rays become duller with every second.

5

The Earth Journey

Strange electro-encephalograms appear on monitors. The humans begin to perceive themselves as a whole and a part of the infinite world, defying gravity like acrobats who never get dizzy. They are puffy cumulus and nimbus clouds, molecules pouring out to flood rivers and deserts. They are trees whose thirsty roots burrow deep into the ground so that their welcoming branches and foliage can thrive. They are plankton in oceans teeming with life, where primitive bacteria craving sugar first created oxygen through countless ploys. Thus begins their initiatory journey.

They experience neither cold nor hunger; they barely even resemble ordinary humans. Whirling like spirals through prehistory, they head for the Earth's red-hot core, penetrating incandescent matter, traversing sediments, metals, and gases. This Earth tells the story of humanity and how time has unfolded in the Universe. Rocks of the cosmos, stones and rivers, lakes and caves, skeletons and bones of all preceding life forms, human bodies buried or forgotten beside their belongings become remains of the past: it all has meaning to those who can see, listen, touch. Everything is evidence of our history. Buried civilizations once oblivious to others, filled with hatred and contempt, devoted to war, now reveal their secrets, their weapons, their vulnerability.

Even though the humans believe they are all following the same route, each one contemplates with a different heart the Earth that their ancestors populated on every continent. Fruitful or arid, alarming or comforting, every land is part of an ecosystem with a fragile equilibrium.

With their bodies stretched out on the Earth's crust, their souls rise up, first one by one, then by hundreds, thousands, and millions, forming a cloud of white-winged unicorns carried along by the winds of Africa, the birthplace of their forefathers, the first of their species.

They ride across the dusty, arid desert of Namibia, where the last remaining trees will soon be swept away by scorching winds, where oryx with long, sharp horns defy lions, and where seeds slumber until the next rainfall before blossoming. They go further still, toward the sweltering Ndoki jungle inaccessible to human beings, toward the high canopies that are home to parrots and bats. As they approach, all the creatures look up at this emanation of the Divine Breath, which now hovers peacefully above them. It is presence.

From the highest leaves to the depths of the humus, insects are everywhere. The humans can see even the tiniest among them. The insect invasion of Earth has been under way for four hundred million years. A million species dominate the world. Those that flourished were the ones that tended to their young, to their future, because a single rule applies throughout the animal kingdom—and that includes humanity. Even the trees converse and cooperate underground.

Eager to celebrate the life teeming on this odd vessel, the cells within them enthusiastically cry out:

> *O! beloved Earth, let them discover your secrets. You can see that they have opened their eyes and are drinking you up with their souls. Give the mortals a warm welcome; their breath is immortal. Let them gorge themselves on you.*

Having spent so much time in ignorance of the life around them, as they skim through the skies the humans discover their planet at long last. Passing over the Pinnacles in Western Australia affords them a commanding view of white dunes, great gray kangaroos, and pink parrots. Almost as soon as they have left the desert region behind them, they head for the verdant, snow-capped mountains and frozen lakes of New Zealand, flying above fjords and dolphins. In Antarctica, a world of once-eternal ice and snow that global warming has been steadily shrinking, elephant seals and black penguins with white bellies have learnt to adapt. Further north, by swimming in tightly-knit schools that seem to form a single creature, anchovies outwit exasperated predatory sea lions. Female octopuses, those solitary, crepuscular, carnivorous creatures, lay up to a hundred thousand eggs each and attach them in clusters to the underside of submerged rocks, cleaning the eggs with their tentacles, providing constant maternal care, only to die once the eggs have hatched. Yet these orphans at birth already know all the adaptive strategies of their species.

The travelers reach the shores of the Galapagos Islands. Not far from the now-replenished coral formations, iguanas, penguins, sea lions, and turtles frolic, indifferent to human presence. It's the same for the Komodo dragons—those giant lizards living on the Indonesian isle of Komodo, where coral reefs provide a protective barrier for dazzlingly colorful fish.

After the hot, rainy, verdant Equator where life shows such beauty and diversity, the travelers look down in fascination on the Pantanal floodplains of Brazil. Submerged during the rainy season, this modern-day Noah's Ark is home to tapirs and giant ants, wolves and monkeys, capybaras and jaguars, along with cowboys herding zebus. The winds drive the humans toward the Angel Falls and tabletop mountains of Venezuela, where wind erosion has produced strange, towering rock formations that rise up out of the jungle. Well before humanity emerged, monsters left their shadows here.

What a marvel to see everything, to discover all the wonders on Earth! Such great migrants as monarch butterflies, with their distinctive orange, black, and white wings, flying instinctively southward toward the warmer climate of Mexico, covering thousands of miles to escape the North American winters. Millions of them travel in clusters until they can form colonies in the sacred fir trees of miraculous forests. Only their descendants—several generations removed—will make the trip back north to reach their exact starting-point. Consciousness is everywhere.

The humans fly over the gorges of North Carolina, where white flowers with delicate corollas grow. They follow the mountain ranges up to Banff National Park, where caribou graze in conifer forests and among ice fields bordering glaciers. They witness the most breathtaking Northern Lights over Ungava Bay and dance above caribou and polar bears in Alaska to the rhythm of the northern magnetic pole. Below them, reindeer and migratory birds cross frozen plains, gambling with their own lives (and deaths).

On the ice field swept by winds of close to one hundred miles an hour, merry seals endowed with incomparably rich milk play with their young before plunging them into deep water, where all too often killer whales await with gaping jaws. But while life is transient everywhere on the planet, it still prevails, whether in dark, icy seas or alongside lava flows.

The humans survey the expanse formed by China, flying over Yunnan Province, where snow leopards hunt and people cross rivers using cable slides, and continue onward to the high plateaus of Tibet, where nomads herd kiang wild asses, yaks, and goats to more plentiful vegetation.

This first journey to the heart of the Earth fills the humans with wonder. The current civilization will not be the last one after all. Disaster was not brought on by striking cells, but by souls gone astray.

The living embrace. So our cells are demanding a different lifestyle? The Elders, previously self-absorbed and self-centered, are invigorated by newfound energy, feel able to accept their own white hair, and allow tenderness to enter their wrinkled faces. People of power are suddenly struck by their own vulnerability, the first stage in the journey along the road to humanity. Children will no longer have to pay for the folly of adults who are haunted by arrogance, revenge, and suffering.

Serenity. Factories are closed, computers turned off. All labor is disalienated on the same day. The humans share sacred time. They meditate, finally entering the cosmic rhythm.

6

The Cosmic Journey

The music of cells, humans, and universes reaches the humans, a soothing music resonating in them at long last, with previously unfamiliar sounds that create perfect harmony. There is no longer any separation between them, between matter, movement, and its vibrations.

Although appearing lethargic, the humans set out on a singular journey through time and space, following the dictates of their spirit. It only lasts a few minutes, but it is an inward journey of the liberated soul. No need for thirteen billion years to be awed by the beginning of time, a time without nebulae, stars, or a solar system, in the dense mist of genesis.

They feel as though they were floating again, only to be propelled into the immensity of their galaxy—just one galaxy among billions of others. The Universe, composed of countless explosions, has also settled within them. Born out of this past, the humans now awaken to cosmic time and memory. Of course, they were already there at the beginning of the worlds that followed the birth of the planets, stars, and superclusters, in the stormy fire of the primordial elements that set the stage for life, for their lives. A power is contained within them. They ride along on nascent, shooting, and dying stars, as conscious energy above the receding Universe. They no longer fear the unfathomable, no longer ask why they are not immortal, that unanswerable, unsettling question that once drove them to despair.

Leaving their planet behind, the humans look on in amazement at innumerable comets with ice floe tails. They soon sense they are being increasingly sucked in by a time–space continuum that only exists in relation to matter. Their next encounters are with terrestrial planets like Mars the Red, a few light–minutes away, as well as with gas giants like Neptune. They are surprised to discover that planets "pair up": each sun has its own planet in this infinite, constantly expanding space.

They exit the Milky Way, moving on toward the giant stars of former worlds that are burning up like furnaces, clouds of dust and gases in which supernovae explode and solar systems are born.

At this point, the humans are traveling faster than the speed of light, edging closer to the divine within themselves. The Divine Breath fills and surrounds them:

> *Have no fear! Come to me; I am waiting for you. We can meet again. Light of my long days, I offer you desire in the place of fear. Ride with me. There is no power in fear; there is tremendous power in desire.*

As brave as in the days of their forefathers, feeling freer than ever, the humans approach the primordial galaxies. They perceive them to be receding inexorably into a world with no horizon, with no time other than eternity, a world that unfolds before their very eyes—eyes now sensitive to light energies—resembling dark shadows seen from the Earth. They skim past suns a thousand times larger than the Earth's; they pass through infernos at two thousand five hundred degrees and icy comets, without getting burned.

They marvel at strangely-shaped stars and galaxies: stretched-out, flat, elliptical, looping, arc-shaped. What they see defies the theories of the most brilliant scientists, who aspire to eminence, to explaining the mystery of life and its origins, to discovering whether we are the only ones living on the billions of planets that make up the Universe, and to capturing the invisible X- and gamma rays emanating from stars. They long to develop theories of how other universes reproduce endlessly like colliding bubbles, or to be the one who finally reconciles the as yet irreconcilable postulates of the infinitely large and the infinitely small.

Once again, they detect the Divine Breath singing within them:

> *There is the time of the infinitely small, the incommensurable eternity of the life principle underpinning the connection in the dance of atoms, which scoffs at all theory. And there is the time of the infinitely large, which the sorcerer's apprentice continues to measure in millions of light–years in an attempt to forget that he himself won't be there forever, that the arrow of time is traveling relentlessly forward. But you, child of the Universe, are part of all times, ever since time began.*

Mystery. For four billion years, the dust of the Universe has been floating, separating, swirling, colliding, plunging into the ocean of a planet that has barely cooled off from its boiling-hot metals. And there, in those turbulent waters, in those seas brimming with life, in that luminous matrix, the humans retrace the path taken by the particles of the Universe, as the molecules of stars join in the fertility game, with one cell creating another in a never-ending process. Identical yet infinite, they multiply; reproduction gives way to procreation. A momentous upheaval comes to light: by attraction or out of love, two structures fuse together and take millions of years to procreate. Millions of years go by until she and he can produce chains of proteins unlike any other, until the "I" can emerge. An "I" that is unique, yet infinitely similar to all that is.

A few mutations and a cascade of genes are what determined its greatness and singularity, what inaugurated its quest for meaning, and what gave expression to its primal and final melancholy: a melody with tragic undertones that denies death.

The lifetime of humanity is one with the time of its planet and its perpetuity through the ages. All links in the chain have their part to play. It is up to each generation to say no to apocalypse.

Their eyes wide open with astonishment, the humans now contemplate the world of over thirteen billion years ago with its red hue, and go through the dark mass that preceded the formation of the solar systems, galaxies, and planets. They enter the cosmic soup, explode with joy three seconds after the Big Bang—Oh!—they weigh the Universe, measure the mass of stars and galaxies, decode the immensity of dark energy, the matter that is the real substratum of the Universe but that remains invisible because it absorbs light, an expanding matrix in limitless space.

7

The Quantum Journey

Their next trip takes them to the heart of the infinitely small, where nothing is quite what it seems. The humans thus move closer to where the Universe is nothing but a speck.

This marks the climax of their initiation. The same principles underlying the cosmos, the Earth, species, and neurons are contained, hologram-like, in humans. The relational forces within atoms are also the key to survival for animals, who depend on the quality of bonds between neighbors, clans, families, couples, nature, and self. The instinct for preservation of oneself and one's descendants is omnipresent, with the various species outdoing each other in cunning, resourcefulness, and versatility. Some roll over and play dead, others disguise themselves as females or as plants, still others use poison or resort to allurement, but in every case the purpose is to maintain the links in an eternal chain. At all levels, in every animal and plant, awareness, strategic ingenuity, and energy vibration are marshaled to ensure survival in even the most hostile of environments. Even in an apparent void, the relational forces are at work.

By this point in the initiation process, the proteins and loquacious molecules can no longer contain themselves. Positively itching to explain at least one side of the wondrous, they blurt out in their subliminal language:

These energy waves affect the living right down to the minutiae of their existence, bombarding them with light neutrinos, whose signals our countless cells continually process, decoding them in accordance with each individual's personal makeup. Humanity resonates with the Universe, engaging in perpetual dialog with its environment. Each person has his or her own energy signature, an individual frequency, with unique receptors for capturing the world, making him or her different from all others. Differentiated yet unified, the humans together compose a universal symphony, one that affects the world in turn.

The humans take in the message that everything is relationship and interdependence in support of life. The Divine Breath gives way to the light that they now see:

Matter is not solid—neither stone, nor iron, nor you. The Universe is an immense void in which stray atoms connect up through a colossal force. But to be able to touch yourselves, to stand upright, to relate to your surroundings, and simply to be in the world, you have developed the illusion of hardness, the misconception that matter is material. What you think you are seeing, what you have the impression you are holding, veils the inner power that gives cohesion to the whole.

Knowledge comes to the humans, as to a bride whose veils are lifted. Drawn in by the expanding forces, their souls get carried every which way by the irregular waves of the Universe. They reach the oldest spiral galaxy,

BX442, whose red light traveled more than eleven billion years before being reflected in the most accurate telescopes. The clouds open up, and photons rain down, covering the voyagers with the dust of the firmament, traversing them, energizing them with lights that have reached them in just a few seconds, and with other energies that have taken billions of years to do so. They have ceased to feel cold. They are incandescent.

Atoms meet up, magnetic fields converse, lights become betrothed, and weddings last an instant, reminding us that we are cosmic radiation, a diaphanous tapestry of sacred music that has filled worlds for billions of years.

In solemn meditation, each person now understands him- or herself to be a galaxy encompassing billions of tiny protective universes, evolving on a planet that is itself part of billions of other astonishing universes. The humans have stopped looking elsewhere for what they already possess in abundance within themselves. No other planet could satisfy their longing. Their molecules will inhabit the Earth eternally. The Earth belongs to all species, nourishing and healing, holding back and giving forth. The Universe is both woman and mother; she is also my fellow being.

The humans return at this point to their loving Earth, as to a neglected lover. The Universe is the matrix, the womb from which they sprang, and the Earth is their adoptive mother.

I want to hold you, dear Earth, in my galactic arms. My brothers and sisters are here, in the single home formed by the Universe. It is not what you summon to yourself that matters, but what summons you. Today, it is life itself which finds fulfillment through relations.

It takes the humans just a few moments to come within range of the Earth. As they fly overhead, successive waves of warm air created by winds sway the palm trees below. The humans soon alight, their bodies ready at last to embrace the miracle of life. They shall belong to a generation of prophets with a newfound belief in humanity. They shall partake of the tree of knowledge, using it for the first time not to destroy, but to build, to fertilize, to make the Earth fruitful.

The equation of brotherhood shall be solved. As children of the Universe, as sisters and brothers in light and genetics, human beings are all cousins. There shall be a planetary Sabbath, a day of brotherhood each year that no one will dishonor: no rape or abduction, no pilfering of people, animals, or things.

The humans rise with the wind. Standing tall. What future awaits them?

8
War or Peace?
The Dilemma of Procreation

The question is left hanging. Spermato and Ovula, the two cells of procreation, adamantly insist on going through with the fourth walkout, arguing that they too have the right to strike.

The existence of a sperm cell, it must be said, boils down to a heroic struggle to win out over millions of his kind who want just as badly to fertilize the same egg, in a race for life like no other. That quest for power—inherent in maleness since the dawn of time—is all about conquering the One who awaits the fittest among them, watchful and calm. They fulfill and complete one another.

Spermato and Ovula are determined to get their point across:

> *The aim of our union, the ultimate struggle for the love of life, is to create a humanity with a sense of the sacred. Each person encapsulates millions of years of history, bearing the trace of eternity in his or her immortal molecules. You are the only beings in the Universe to experience cosmic, earthly, and human time all in one. Yet you neglect that privilege.*

With their craving for instant gratification, the humans were indeed unaware that they lived in several worlds and several kinds of time. Or that their lives were in essence defined by relationship, interdependence, change, and diversity. More importantly, they had failed to realize that everything is meaning.

Neither Spermato nor Ovula want to call off the last strike. If such awareness is doomed to disappear anyway, why not get it over with right now? Panic-stricken animals and plants with a long-standing attachment to humans sense that they must once again take resolute action:

> *The Earth without humanity? So we will have to wait for millions of years before we hear a child laugh or weep, before one of them touches us and praises our beauty? Spermato, Ovula, are you insane enough to do such irreparable damage? How can we tell the humans that we love them, that they are not facing this alone?*

The vegetable and animal kingdoms manage to sway Spermato and Ovula. Joining forces, they perform the ballet of life. In a dance of diaphanous wings, billions of foraging and coleopterous insects dart out and zero in on soft, opulent flowers, with their innumerable fragrances and colors. Wooing, copulating—the springtime of life gets under way.

Eager for reproduction on Earth, the flowers display tantalizing corollas designed to attract birds and insects. Taking advantage of their close ties to a number of these fauna, flowers vie with one another, competing with fragrance, color, and form. Orchids in Costa Rica seductively flaunt their exotic gowns. Tropical Heliconias exude large quantities of nectar that soon drive hummingbirds wild. Cactuses in Arizona harboring fragile seeds offer the succulent pleasure of their one-day flowers to bats, birds, insects, and tortoises, who then carry those precious seeds far and wide.

The wind rises, in Europe pollinating pine and ash trees; in southeastern China, pollinating pyramid-shaped ginkgos. Ferns, those ancient elders, embark upon propagation. White nettles offer insects their five petals, while cocoa trees in the Ivory Coast proffer lovely pink flowers. In the Arizona desert, moths pay court to yuccas.

Ah, when the sweet smell of pheromones joins hands with love! exult Spermato and Ovula.

Along the shores of the Atlantic, sea horses gingerly kiss. On the Sunda Islands, freshwater harlequin fish adopt orange and purple hues for the spawning season. King cobras in Burma prepare for the mating season, leaving their venom behind. In Bolivia, male nandus assist their mates in caring for their young, whom they never abandon. Winged termites indulge in merry nuptial dances.

Faithful, gracious cranes, long necks protruding from white bellies puffed up with pleasure, initiate their amorous duet on grassy shores, wiggling their black bottoms in rhythm over clear waters. In the ocean of sand that forms the world's oldest desert in Namibia, a chameleon sets out on a lone journey that will lead him to his nuptials with a female. After a brief embrace, the lovers separate, fully gratified. The species ensures its survival amid oryx and zebra skeletons.

In oceans the world over, humpback whales sing their joyous mating songs, shooting out great plumes of water, vigorously slapping their impressive tail flukes. The females draw males with their boisterous dance, pitting them against one another, before accepting the advances of the fittest. Their young are born weighing a ton and a half, and are nursed for six full months. Flying fish spread their wing-like fins and glide over the water's surface, diving back to lay eggs on floating palm fronds, where the males will fertilize them. Under the weight of the eggs, the fronds may even sink to the depths.

The humans see the future and detect other colors. Plants everywhere, deriving their energy from blue and red light waves ordinarily invisible to humans, emerge from the shadows, stretching their photoreceptors toward the light. Fast-flying hawk moths cross entire countries to inform endangered butterfly species that the humans will not be inflicting pesticides on them any more. Horse flies make their journeys quietly, no longer biting people or livestock. In the jungles of Sri Lanka, peacocks stop displaying their fans to impress their lady–loves, and with uncharacteristic humility, win them over immediately.

Elephant seals give birth once again on the shores of Mexico, now that they have recovered their trust in humanity. Timid llamas on the plateaus of Patagonia, elks in Poland, bison in Nevada, and miniature goats in Southeast Asia have all ceased to fear our species.

In their water of life, Spermato and Ovula embrace.

> *Come to me, my sovereign queen. We shall have seasons and harvests again.*

She calmly lets him penetrate her kingdom like a victor:

And our children will people the Earth without fear.

The procreation strike will not take place. The season of loving gives way to the age of love.

Now resonating with life, the humans awaken to the legacy of self-hatred shifted onto others since time immemorial: the insatiable appetite for power of the power-hungry, fratricidal religions, turf warfare, eternal scapegoating. In every nation, every community, wise figures speak up to apologize for the acts of past generations. Reconciliation becomes a worldwide reality; minorities gain recognition and can at last live in peace. After millennia of tears, sons will no longer have to pay for their fathers' actions. At last they weep, heart pressed up against heart. War will be no more. Tender will be the Earth.

9

The Time of Women
Reaching Out to Men

A woman rises to her feet:

> *We have been strangers to ourselves for too long, waiting to come into ourselves.*

Another, standing just as tall, takes the floor and says:

> *At the end of the road, Life, there is just you and me. At the end of the road, I encounter myself, discover myself, and love you filled with me. Stretching back four billion years, I have been waiting to carry a child and then disappear. To be, countless and fearless. Daughter, sister, wife, mother, woman, lover. Four billion years to experience today the racing of my heart, galloping along the beach toward encounter. Charting my own path at last.*

A woman beside her speaks up in turn:

> *On the tip of your lips, I taste life, solitude, and excess. In your arms, I cradle sorrow. My gaze upon you bears witness to your mystery, containing amazement and laughter, arms and teeth, limbs and desires, tears and inebriation, stars and abundance.*

— Come, Life, says another woman. My body is your dwelling. Come quench your thirst, flow through me. I am a source, torrent, and volcano, I am a cave and a plain. Replete! I am a tiger, elephant, and unicorn, I am a firefly, primate, and citizen.

Women, who suffer more than anyone from the death of a child, from human solitude and the melancholy of Eden, rise to their feet. One takes her place at the end of the Earth, then ten, a hundred, a thousand, and then millions. Four billion women, with hands held out toward men and children, serene and sovereign at last:

At the heart of the world is relationship. If we fear to encounter others, it is because we do not realize that we are greater than ourselves. If we fear strangers, it is because we are in exile from ourselves. If we squander our Earth, it is because we have forgotten our universal heritage. If we make war, it is because we are ignorant of the joys of love. Before parting, let us travel down the road together.

The way is open. Men and their children respond to this invitation, first one by one, then by the hundreds of thousands. Sixteen billion palms—smooth and coarse, slight and bulky, in every tone and color—recognize one another. Hands that faintly touch each other at first soon clasp together, clasping hearts at the same time. Eyes glisten with crystalline tears. A chain of human beings spanning all continents links together. Tectonic plates once smashed up against each other, forming separate continents. They now join together in an imperceptible serenade.

A dip into human time. All are Africans, sisters and cousins on a quest for new horizons and more hospitable lands. Across continents and countries, forming tribes and nations, they are unique and alike in their fears and aspirations. From the abyss to the summit, they finally know the way.

Eight billion men and women become Humans. Homo Filia, the being of connection, is born. They all hear themselves calmly articulate a "No" to the power-thirsty rulers. The Earth is their garden, theirs to cultivate. A breeze gently caresses their illuminated faces. In each of them, these words reverberate:

> *Listen, O life. Come, for we will set out to find you wherever you be.*

Women program their daughters for a future longer than that of their sons, perhaps hopeful that such longevity will bring an awakening that can enlighten humanity at last. At this point, they all rise to their feet.

> *Do you hear the voice, do you see the signs wherever you look? sing the matriarchs. What do you know of this dance of life that you are so afraid of discovering? What fears trouble this heart that is yours, yet unknown to you? Do you fail to receive all the messages assailing you saying, "Once upon a time there was Singularity"—and that this refers to YOU? Child of the Big Bang, the primordial matrix in which your soul bathes, take delight in giving birth. Cousin of gases and dust, of rocks and oceans, of sharks and canaries, of lions and bees, relative of all Earth dwellers, illustrious before returning to dust. Let me tell you that you are a child of this*

Earth. Come dance to the eternal hymn of life, together with your amnesiac brothers.

You, I, and the Other share the same journey, an inescapable destiny that it would be vain to resist. We were dwellers of the Earth, but what about tomorrow?

The Universe will be our soul when it is relieved of our bodies, as we shall be its soul once again. At that point, we shall return home, to where we originated. Beyond doors and walls, lifting our veils, opening our eyes wide, heading for the invisible as it manifests itself.

Tell the humans, who invented separation, irreconcilable differences, impossibility, antithesis, and stalemate; tell those who swear by the concepts of either/or, denoting opposition, and but, which excludes, annihilates, oppresses, and victimizes, while forgetting and, which includes, heals, and reunites; tell all of them about the foundations upon which life and the Universe rest. About the communion of atoms, repelling and attracting each other; about the dance of jumping genes that encode universal information and reinvent themselves; about suns coupling with their moons. Beyond the bounds of fear lies relationship; within the bounds of fear lie hatred and isolation.

Tell them that life means bonds and connections, that it exalts interdependence and autonomy. Body and spirit and soul are one. We are moved by reason and emotion, by determinism and free will. Community and ego can coexist. The irreconcilable can be reconciled.

The time of and will come, the time of children of Singularity, in which "Us against Them, the ones to oppress," will be no more. Let us say "No" to the fear of encounter.

The girls stand up, their hands outstretched to the men and children. The Whole has always been contained within them.

> *Come to me, all of you. I want to live, laugh, dance and smile with you. Come, my beloved, my transparent wings, my transient husband, my temporary mooring, my breath, my last sigh. Come, for wherever you are, I will go to find you.*
>
> *Now that my flesh-and-blood mother has departed on her long journey, I shall also return to the asteroid dust I once was. Did she not tell me that I would be a star once again, she who dreamt of stardom?*
>
> *The unfathomable firmament will be our cloak, and the galaxies, overjoyed at being together, will provide our bedding.*
>
> *And now, Mystery, come to me and touch me. I am calling to you again. Come before my eternity.*

Acknowledgements

I would like to thank my family, to whom this book is dedicated: my children Karen and Daniel, my grandchildren David, Naomie, Ilan, Olivia, Lily, Avi, and Elijah, my sisters Jacqueline and Gigi, my brothers David, Jose, and Charly, their children, and our future descendants.

I also owe a great debt of gratitude to Albert Jacquard, Jacques Salomé, Oprah Winfrey, Michel Blanc, Lison Benarroch, Francine Léger, Martha Lawee, LeeAnn Pollack, Guillaume de Lacoste Lareymondie, Larry Cohen, Michèle Dernis, Christine Fain, and Alexandra Fain for their contributions.

Finally, my thanks go to all my friends.

V. A.

Bibliography

Ancelin Schützenberger, Anne. *Aïe, mes aïeux.* Paris: Desclée de Brouwer, 1993.

Asimov, Isaac. *Asimov's New Guide to Science.* New York: Basic Books, 1984.

Brizendine, Louann, M.D. *The Female Brain.* New York: Harmony, 2007.

Brizendine, Louann, M.D. *The Male Brain,* New York: Harmony, 2011.

Buber, Martin, *The Way of Man.* London: Routledge, 2002.

Casse, Michel. *Trous noirs en pleine lumière.* Paris: Odile Jacob, 2009.

Collective work. *La recherche en neurobiologie.* Paris: Point Seuil, 1988.

Furst, Charles. *Origins of the Mind: Mind-Brain Connections.* Upper Saddle River: Prentice Hall, 1979.

Girard, René. *Things Hidden since the Foundation of the World.* Redwood City: Stanford University Press, 1987.

Goleman, Daniel. *Social Intelligence.* New York: Bantam, 2007.

Gusdorf, Georges. *Speaking (La Parole)*. Evanston: Northwestern University Press, 1979.

Jacquard, Albert. *In Praise of Difference: Genetics and Human Affairs*. New York: Columbia University Press, 1984.

Jacquard, Albert. *Moi et les autres*. Paris: Le Seuil, 1983.

Juan de Mendoza, Jean-Louis. *Cerveau gauche, cerveau droit*, Paris: Flammarion, 1995.

Kurzweil, Ray. *The Singularity Is Near*. New York: Penguin, 2006.

Laborit, Henri. *Éloge de la fuite*. Paris: Robert Laffont, 1985.

Laborit, Henri. *La légende des comportements*. Paris: Flammarion, 1994.

Lipton, Bruce H. *The Biology of Belief*. Carslbad, CA: Hay House, 2007.

Melzack, Ronald and Wall, Patrick David. *The Challenge of Pain*. London: Penguin Global, 2004.

Navarro, Joe and Poynter, Toni Sciarra. *Louder Than Words*, New York: William Morrow, 2011.

Neuberg, Marc, Ewald, François, Hirsh, Emmanuel and Godard, Olivier. *Qu'est-ce qu'être responsable?* Paris: Edition de Sciences Humaines, 1997.

Novák, František Antonín and Cuzin, Michel. *Encyclopédie illustrée du monde vegetal*. Paris: Gründ, 1971.

Olivier, Christiane. *Les enfants de Jocaste*. Paris: Denoël, 2011.

Olson, Steve. *Mapping Human History: Discovering the Past Through Our Genes.* Boston: Houghton Mifflin Harcourt, 2002.

Reeves, Hubert. *Patience dans l'azur.* Paris: Le Seuil, 1981.

Rosenzweig, Marc and Leiman, Arnold. *Physiological Psychology.* New York: McGraw-Hill, 1989.

Salomé, Jacques. *Pour ne plus vivre sur la Planète Taire.* Paris: Albin Michel, 2004.

Salomé, Jacques. *Parle-moi– j'ai des choses à te dire.* Paris: Editions de l'Homme, 2004.

Schroeder, Gerald. *The Science of God.* New York: Free Press, 2009.

Solomon, Eldra Pearl and Davis, P. William. *Human Anatomy and Physiology.* Philadelphia: Saunders College Publishing, 1983.

Stanek, V. J. *The Pictorial History of the Animal Kingdom.* Worthing: Littlehampton Book Services Ltd., 1968.

Stent, Gunther S. *The Coming of the Golden Age. A View of the End of Progress.* Boston: Natural History Press, 1969.

Turchet, Philippe. *Le langage universel du corps.* Paris: Editions de l'Homme, 2009.

Vedantam, Shankar. *The Hidden Brain.* New York: Random House, 2010.

Weinberg, Steven. *The First Three Minutes: A Modern View of the Origin of the Universe.* New York: Basic Books, 1993.

Zavalloni, Marisa et Louis-Guérin, Christiane. *Identité sociale et conscience*. Montreal: Presses de l'Université de Montréal, 1984.

Periodicals

- *Alternatives internationales*, "Les guerres des matières premières", July 2012.

- *Discover*, "Technology on the Brain," September 2013.

- *Discover*, "How to Cure Everything," October 2011.

- *Discover*, "Evolution Next Stage," March 2012.

- *Discover*, "Extreme Universe," Winter 2010.

- *Discover*, "The Brain, Mind and Machine," Spring 2012.

- *Les cahiers Science et vie*, "L'homme et la machine," October 2012

- *How It Works*, no. 42.

- *Sciences et avenir*, "Les origines de nos croyances," Special Edition no. 173.

- *Sciences et avenir*, "Les physiciens ont capturé l'antimatière," no. 767.

- *Sciences et avenir*, "La science sait lire dans nos pensées," no. 1098.

- *Sciences et avenir*, "Les nouveaux mystères du sexe," March 2012.

- *Sciences et avenir*, "Grandes avancées et nouveaux espoirs," no. 1120.

- *Sciences et avenir*, "Cosmos, les ultimes défis," June 2012.

- *Sciences et avenir*, "Incroyables neurones," October 2012.

- *Science et vie*, October 2012.

- *Science et vie*, "Temps, matière et espace," Special Edition, September 2012.

- *Science et vie*, "Spécial 100 ans," April 2013.

- *Science et vie*, Special Edition, 2013.

- *Scientific American*, "The Brain's Dark Energy," March 2010.

- *Scientific American*, "Food," September 2013.

- *Scientific American*, "Evolution of Creativity," March 2013.

- *Scientific American*, "How Minds Bounce Back," March 2011.

- *Scientific American*, "The Real Sexual Evolution," January 2011.

- *Scientific Mind*, "Shh, I Am Getting Smarter," September 2008.

DVDs

- *Planet Earth*, BBC Series, narrated by David Attenborough

- *Life*, BBC Series, narrated by Oprah Winfrey

Made in the USA
Charleston, SC
09 February 2015